Rodney A. Gayer

The Tectonic Evolution of the Caledonide-Appalachian Orogen

International Monograph Series
on Interdisciplinary
Earth Science Research and Applications

Editor
Andreas Vogel, Berlin

Editorial Advisory Board
D. E. Ajakaiye, Zaria/Nigeria
E. Banda, Zurich/Switzerland
R. Dmowska, Cambridge, MA/USA
R. A. Gayer, Cardiff/UK
R. Greiling, Mainz/FRG
F. Hórvath, Budapest/Hungary
E. Mantovani, Siena/Italy
E. Milanowsky, Moscow/USSR
Ö. Öztunali, Istanbul/Turkey
I. Ramberg, Oslo/Norway
A. M. C. Sengör, Istanbul/Turkey
D. H. Tarling, Newcastle upon Tyne/UK
D. A. Warnke, Hayward, CA/USA

Rodney A. Gayer (Ed.)

The Tectonic Evolution of the Caledonide-Appalachian Orogen

Friedr. Vieweg & Sohn   Braunschweig/Wiesbaden

CIP-Kurztitelaufnahme der Deutschen Bibliothek

The **tectonic evolution of the Caledonide appalachian Orogen** / Rodney A. Gayer (ed.). — Braunschweig; Wiesbaden: Vieweg, 1985.
  (Earth evolution sciences)
  ISBN 3-528-08596-7

NE: Gayer, Rodney A. [Hrsg.]

All rights reserved
© Friedr. Vieweg & Sohn Verlagsgesellschaft mbH, Braunschweig 1985
No part of this publication may be reproduced, stored in a retrieval system or transmitted, mechanical, photocopying or otherwise, without prior permission of the copyright holder.

Produced by Lengericher Handelsdruckerei, Lengerich
Printed in Germany

ISBN 3-528-08596-7

Contents

Editorial
R. A. Gayer 1

British Caledonian Terranes
W. Gibbons/R. A. Gayer 3

Relationships of Cambrian-Ordovician Faunas in the Caledonide-Appalachian Region, with Particular Reference to Trilobites
W. T. Dean 17

The North American Appalachian Orogen
R. D. Hatcher, Jr. 48

Low-Grade Metamorphism in the Paratectonic Caledonides of the British Isles
R. E. Bevins/G. J. H. Oliver/L. J. Thomas 57

A Review of Caledonian Volcanicity in the British Isles and Scandinavia
R. E. Bevins/C. J. Stillman/H. Furnes 80

The Role of Thrusting in the Scandinavian Caledonides
J. R. Hossack 97

A Major Stratigraphical and Metamorphic Inversion in the Upper Allochthon of the Scandinavian Caledonides
R. Mason 117

Caledonide-Appalachian Tectonic Analysis and Evolution of Related Oceans
A. J. Barker/R. A. Gayer 126

Book Reviews 167

Report on the 5th International Conference on Basement Tectonics .. 192

Editorial

The Caledonide-Appalachian orogen is a fold and thrust mountain belt formed between 700 and 400 million years ago. Prior to Mesozoic/Cenozoic spreading of the Atlantic Ocean it formed a continuous linear chain extending some 10,000 km from Svalbard, in the Arctic, to the Gulf of Mexico in the south and with an average width of about 1000 km.

The geology of the Caledonide-Appalachian orogen is probably the most intensively studied of all mountain chains, and yet its origins and evolution are still highly controversial. Interest in the orogen has been heightened in recent years by a vast amount of new information arising from work in connection with I.G.C.P. project no. 27 — the Caledonide Orogen. This information has emphasised the complexity of the chain, so that, although there is little doubt that the orogen was developed in association with an evolving ocean, the Iapetus Ocean, the history of this evolution is extremely complex. A simple plate tectonic model involving the opening and subsequent closure of Iapetus can no longer be maintained.

The new data have revealed the diachronous and polyphase nature of the deformation and metamorphism, the presence of many tectonically emplaced ophiolite fragments representing remnants of oceanic lithosphere, and the occurrence of numerous volcanic sequences variously attributed to initial lithospheric stretching, to arc complexes, and to back arc basins. New studies of Early Palaeozoic faunas have helped clarify the two continental shelf margins of Iapetus but have also demonstrated intermediate and endemic faunas characteristic of neither margin.

It is clear that, in addition to modelling the Caledonide-Appalachian orogen on present plate boundaries, it is necessary to recognise the importance of major fault contacts. As with the Western Cordillera of North America, where stratigraphic, structural and palaeomagnetic studies have demonstrated the presence of a large number of suspect terranes, juxtaposed by major strike-slip displacements, so the Caledonide-Appalachian orogen is best investigated by terrane analysis. Each terrane is regarded as fault displaced unless a definite connection between adjacent terranes can be established.

The articles presented in this thematic issue have been carefully selected to give a wide coverage of the orogen, reviewing the important facets of its evolution in terms of plate tectonics and suspect terrane analysis. Such an approach, of necessity, covers a wide spectrum of earth science disciplines but every effort has been made to integrate the individual articles into a general framework. Thus the first article on British Caledonian terranes lays the foundations for the main reviews and the final article presents a coordinated tectonic model for the evolution of the belt, integrating the material from the individual contributions.

The accounts presented in this issue represent an up to date overview of the evolution of one of the most intriguing of orogenic belts. It is hoped that they will be of interest to all students of Caledonide-Appalachian geology, whether they are coming to the orogen for the first time or continuing their research to reveal even more of its fascinating history.

University College Cardiff *Rodney Gayer*
March 1984

British Caledonian Terranes

W. Gibbons/R. A. Gayer
Department of Geology, University College, P.O. Box 78, Cardiff CF1 1XL, U.K.

Key Words

British Caledonides
Suspect terranes
Transcurrent faults
Iapetus suture

Abstract

The British Caledonides are divisible into at least nine ENE-WSW trending linear terranes most of which are < 120 km wide. From north to south these are: Northwest Foreland; Northern Highlands; Grampian; Midland Valley; Southern Uplands; Lakesman; Monian; Southern Britain; Northern Armorican. Where exposed, the boundaries between these terranes are major crustal lineaments (thrusts or steep faults) and each terrain exhibits its own distinctive pre-Devonian geological history. Given these differences and the importance of strike-slip movements along modern active plate margins, it is postulated that major transcurrent fault movements have occurred between the majority of those British Caledonian terranes bounded by steep faults. The transcurrent movements were probably oriented approximately ENE-WSW and have remained undetected by palaeomagnetic data. The process of orogenic terrane accretion began at least as early as the latest Precambrian, with important phases of activity occuring during the early Ordovician (Grampian) and the late Silurian (end-Caledonian).

There are many broad similarities between the British Caledonides and Appalachian terranes, particularly in the approximate timings of orogenic events, although most of the British terranes have no direct correlatives with those in the Appalachians. By contrast, the Scandinavian Caledonides are dominated by thrust tectonics and, unlike the British and Appalachian areas, are not segmented into suspect terranes mostly bounded by steep faults. A major phase of early Ordovician orogenic activity in N Scandinavia (Finnmarkian) is approximately coeval with the Taconic-Grampian in the Appalachians and Britain, but occurred on the opposite side of the Iapetus Ocean. The main collisional event in the Scandinavian Caledonides (Scandian) occurred in the middle Silurian before the relatively mild closure of Iapetus in the British Caledonides (end-Caledonian event: late Silurian) and the Appalachians (Acadian event: Devonian).

Introduction

The importance of strike slip movement to orogenic belts is now widely appreciated and the concept is incorporated into virtually all models which attempt to reconstruct the history of such regions (eg. Dewey 1982). Most formerly active plate margins will consist of a mosaic of crustal fragments (the 'suspect terrains' of Coney et al. 1980), moved to their relative positions along major, often transcurrent faults. The British Caledonides is considered to be no exception to this general rule, being divisible into at least nine such terranes (Fig. 1). Where exposed, the boundaries between these terranes are major crustal discontinuities with marked contrasts in the geology on either side, and therefore should be considered as 'suspect'. This paper takes the 'guilty until proven innocent' view (c.f. Williams and Hatcher 1982) that major displacements between these terranes are likely to have occurred unless it can be demonstrated otherwise.

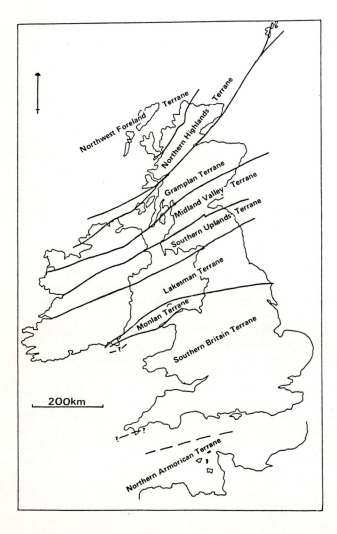

Fig. 1

Geographical location of British Caledonian Terranes.

The position of the concealed boundary between the Sothern Britain and Northern Armorican Terrains is arbitrarily drawn along the English Channel.

It is often difficult, if not impossible, to deduce the amount of movement which took place between terranes, the time of their 'docking', and even the exact position of the major lineaments separating them. The following short review attempts to highlight important similarities and differences between British Caledonian terranes and compares the situation in the British Caledonides with that elsewhere in the Appalachian-Caledonian Orogen.

Northwest Foreland Terrane

A small part of the Northwest foreland to the Caledonian orogen is preserved in a narrow strip some 300 km long in NW Scotland where an ancient gneissic basement (Lewisian) is overlain unconformably by Upper Proterozoic red beds (Torridonian). In NW Scotland granulite facies Lewisian gneisses have yielded Archaean dates ('Scourian' c. 2800 Ma) although in many places these have been reworked under amphibolite facies during the Proterozoic ('Laxfordian' c. 1700 Ma) (Watson 1975). The Torridonian sediments are likewise divisible into an older unit (Stoer Group c. 1100 Ma) and a younger one (Torridon Group c. 1040 Ma) (Smith et al. 1983). Both Lewisian and Torridonian are overlain by a virtually undisturbed Cambro-Ordovician 'Durness' succession of shallow marine sediments containing a fauna of Laurentian affinity (Dean, this volume).

Extensive exposures of this same foreland, dispersed by the later opening of the North Atlantic, are found in NE America, E Canada, NE Greenland and E Svalbard (Swett 1981). However, a pronounced dissimilarity between the NW Scotland palaeomagnetic a.p.w. path and that obtained from North America (Briden et al. 1973, Turnell & Briden 1983) indicates the likelihood of substantial movement of the terranes in NW Britain relative to the main mass of the Laurentian craton.

The Northwest foreland in Britain is bounded to the SE by the Moine Thrust Zone, the age of the main movement on which is poorly defined. Although the latest, brittle thrust movements could be as young as early Devonian (Elliott and Johnson 1980), the main mylonitic shear zones were probably formed sometime between the mid-Ordovician (Llanvirn) and early Silurian (Llandovery) (Higgins 1967; Van Breeman et al. 1979; Parsons and McKirdy 1983).

Northern Highlands Terrane

The metamorphic rocks of the Northern Highlands of Scotland are bounded to the NW by the Moine Thrust Zone and to the SE by the transcurrent Great Glen Fault. Most of this terrane consists of the 'Northern Moines' — a series of metasediments which have suffered both 'Grenvillian' (c. 1000 Ma) and 'Caledonian' (c. 450 Ma) tectonometamorphic events (Brook et al. 1976; Van Breeman et al. 1974, 1979; Brewer et al. 1979). The Moinian has been divided into three main divisions: a central high grade (often migmatitic) belt known as the Glenfinnan Division separates the lower grade Morar (NW) and Loch Eil (SE) Divisions. Several major shear zones have been recognised within the Northern Moines, the most notable being the Sgurr Beagh Slide which separates the Glenfinnan Division from the Morar Division beneath (Piasecki et al. 1981). These shear zones appear to be structurally higher and more 'internal' representatives of the NW verging Caledonian Moine Thrust system. Several dates of c. 750 Ma have been obtained from Moinian pegmatites in this terrane and these have been used to define a 'Morarian' event (Van Breeman et al. 1974; Brook et al. 1977), although whether this was an orogenic event of regional importance is questionable (Powell et al. 1983).

Reworked inliers of Lewisian basement gneisses occur within this terrane, often as allochthonous tectonic slices within the Moines (Rathbone and Harris 1979). A possible SW extension of this terrane may be traced to the island of Inishtrahull off the N coast of Donegal where is exposed the most southerly outcrop of Lewisian gneisses in the British Isles. Whether these outcrops actually represent a continuation of the Northern Highlands Terrane, or whether they form an allochthonous fragment within the Great Glen Fault system remains unproven. The Inishtrahull gneisses have yielded both pre-Grenvillian and lower Palaeozoic RB/Sr and K/Ar dates, which suggest them to have been involved in Caledonian reworking (Bowes and Hopgood 1975; Macintyre et al. 1975). A similar microterrain is exposed to the west of the Walls Boundary Fault in Shetland (Flinn et al. 1979).

Grampian Terrane

This terrane is bounded by major, steep faults: the Great Glen Fault (NW) and the Highland Boundary Fault (SE). Although Moinian rocks similar to those found in the Northern Highlands occur within the Grampians, a major difference between these two terranes is the dominance of Dalradian metamorphic rocks SE of the Great Glen Fault. Lewisian rocks appear in the islands of Islay and Colonsay (Smith and Watson 1983), although it is not clear whether these outcrops belong to the Grampian terrane or are in fact north of a splay from the Great Glen Fault system. Similarly anomalous are three possible outliers of Dalradian rocks just NW of the Great Glen Fault (Garson and Plant 1972; Smith and Watson 1983). The Moines are divisible into a structurally complex, high grade unit (Central Highlands Division) (CHD) and a lower grade unit (Grampian Division) (GD). These have been interpreted as representing a basement-cover sequence. The basement (CHD) has so far yielded minimum metamorphic ages of 730–780 Ma but may have been involved in the c. 1000 Ma Grenvillian event (Piasecki and Van Breeman 1979; Piasecki 1980). The cover (GD) is structurally less complex and separated from the CHD by a Caledonian shear zone (the Grampian Slide). The Grampian Division grades conformably upwards into the lowest Dalradian unit (Appin Group).

It is notable that the main Caledonian deformation and metamorphism took place earlier (> 490 Ma) in the Grampian Terrane than in the Northern Highlands Terrane. This complicates any easy correlation between the two metamorphic sequences. A certain degree of diachroneity from the internal (early) towards the external (late) part of the orogen is to be expected, but a span of > 60 Ma is excessive. Major strike-slip displacement along the Great Glen Fault system during the Silurian, after the movement of the Moine Thrust but before the emplacement of the later Caledonian granites, combined with oblique collision (and therefore diachronous orogeny) could explain these differences (c.f. one alternative suggested by Coward 1983). Although limited late movement has undoubtedly taken place along the Great Glen Fault system, the docking of these two terranes is interpreted to have been effected by a phase of post-collisional transcurrent movement which took place in mid- to late Silurian times.

The Grampian Terrane may be traced NE to eastern Shetland where it is underlain by a Moine basement (Flinn et al. 1972, 1979), and SW to northwest Ireland where most of the metamorphic rocks are again Dalradian. An old Moine-type basement is locally exposed in North Mayo (yielding Grenvillian U/Pb zircon ages) and possibly in the Ox Mountains (yielding a 'Morarian' $^{40}Ar/^{39}Ar$ date). There are many clear stratigraphic and structural similarities between the Irish and Scottish Dalradian (e.g. see Phillips 1981). A notable anomaly however, is the position of the most southerly Dalradian inlier in Ireland, the Connemara massif. The high grade Connemaran rocks, together with the Lower

Palaeozoic, low grade South Mayo Trough immediately to the north, appear to form a small terrane which is allochthonous with respect to the Grampian Terrane i.e. SE of the presumed SW continuation of the Highland Boundary Fault. Explanations provided for the anomalous position of this microterrane emphasise either transcurrent movement (e.g. Dewey 1982) or SE directed major thrusting (Leake et al. 1983).

Another enigmatic microterrain is preserved by the ophiolite nappes of NE Shetland (Flinn et al. 1979). These rocks appear to have been obducted northwestwards over 'Grampian' rocks and now lie close to the probable NE extension of the Great Glen Fault known as the Walls Boundary Fault in Central Shetland. They have no obvious correlatives within the main mass of the Grampian Terrane further SW, and may represent either a fragment of the Iapetus Ocean or oceanic crust once lying NW of the Iapetus and beyond a relatively buoyant crustal unit (e.g. an arc — see Mitchell 1984) which collided with the Laurentian craton to induce the ophiolitic obduction.

Midland Valley Terrane

The pre-Devonian geology of the Midland Valley of Scotland is completely different from that of the Grampian Terrane to the northwest. The Dalradian succession is abruptly terminated by the Highland Boundary Fault south of which the oldest rocks are low grade Ordovician (at Balantrae) to mid-Silurian (e.g. the Pentland Hills). There is inferential evidence for a high-grade granulitic (rather than Dalradian) basement to the Midland Valley Terrane (Upton et al. 1983). This again suggests that major later-Silurian transcurrent movement was responsible for the juxaposition of these terranes. The Highland Boundary Fault contains a number of tectonic slivers of altered Cambrian (or Cambro-Ordovician) ophiolitic rocks known as the Highland Boundary Complex. The age and origin of this Complex remain controversial (Henderson and Robertson 1982; Curry et al. 1982; Ikin 1983) but their restriction to the fault area suggests they may be allochthonous slivers transported laterally along a transcurrent system after the Grampian collisional orogeny. Similar rocks are exposed in W. Ireland along the presumed continuation of the Highland Boundary Fault. These rocks, known as the Deer Park Complex, have been interpreted as representing a dismembered ophiolite occupying a major N dipping shear zone (Ryan et al. 1983). Structural data for the Deer Park Complex suggest a combination of northward directed thrusting with subsequent strike-slip movement (Ryan et al. op. cit.). Mitchell (1984) has interpreted the Highland Boundary Fault as a transpressional, sinistral, oblique-slip fault dipping steeply to the northwest.

In the SW of the Midland Valley of Scotland the Lower Ordovician Ballantrae Complex, overlain unconformably by Llanvirn sediments, has been interpreted as representing oceanic crust obducted prior to the growth of the Southern Uplands 'accretionary prism' during the Llandeilo epoch (Leggett et al. 1979). The Ballantrae rocks lie just northwest of the supposed SW continuation of the Southern Upland Fault (the Stinchar Fault). However, the relationship between the ophiolite and the Midland Valley Terrane is obscured by younger rocks. It remains possible that the Ballantrae succession is an allochthonous, fault bounded fragment within the Southern Upland Fault system.

Southern Uplands Terrane

This terrane comprises the low grade Lower Palaeozoic (Llanvirn to Wenlock) accretionary prism of Leggett et al. 1979. It is bounded to the NW by the Southern Uplands Fault and to the SE by the inferred major lineament of the Iapetus Suture (Phillips et al. 1976) i. e.

the tectonic line representing the junction between what were once opposite sides of the Iapetus Ocean. The terrane extends SW into the Longford Down Massif of Ireland. The apparent existence of continental crust beneath the Southern Uplands raises a difficulty which requires the entire accretionary prism to be allochthonous (Upton et al. 1983).

There are great differences between the Silurian sediments of the Southern Uplands and those exposed as inliers in the Midland Valley (e.g. see Cocks et al. 1971). The inliers of Silurian rocks in the Midland Valley show a shallowing upwards marine clastic sequence of late Llandovery age, passing upwards into Wenlock and Lower Ludlow red beds. By contrast, the Silurian of the Southern Uplands consists of a Lower to Middle Llandovery graptolite shale facies overlain by thick upper Llandovery to Upper Wenlock turbidites. The sedimentary facies, sub-greenschist metamorphism and structural style of the latter accord with their development as an accretionary prism from Llanvirn to Wenlock times (Leggett et al. 1979). This contrasts with the very low grade metamorphism and gentle folding of the Midland Valley inliers. In addition southeasterly derived Silurian conglomerates of the Midland Valley have no known source within the Southern Uplands (Hall et al. 1983). However, like all previous terranes mentioned above, the Southern Uplands Terrane was intruded by a suite of 'Newer Granites' in end Silurian — early Devonian times (Halliday et al. 1979) and overlain by Devonian molasse. On the basis of these mid Silurian differences and early Devonian similarities, a period of late Silurian (Ludlow) transcurrent movement between these two terranes is postulated.

Lakesman Terrane

The inferred position of the Iapetus Suture is generally drawn along the Solway Firth and down through Central Ireland (Phillips et al. 1976), although the suggestion has recently been made that in Ireland the Suture may lie much further north (Leake et al. 1983). The latter suggestion underlines the difficulty of defining the precise whereabouts of what must nevertheless be a major crustal lineament. The Lakesman Terrane includes the Lower Palaeozoic rocks of the Lake District, the Isle of Man and probably the Ingletonian of North Yorkshire. It continues into the Leinster Massif of SE Ireland (e.g. Kennedy 1979). Marine sediments were deposited in this terrane from early Cambrian to late Silurian times (Brück et al. 1979; Moseley 1978), with an important period of calc-alkaline subduction related volcanicity occurring during the Ordovician (Bevins et al. this volume). Any accretionary prism produced during this subduction activity is apparently not now preserved, the volcanic rocks of the arc lying only some 30 km south of the Iapetus Suture. Leggett et al. (1983) have suggested that both the accretionary prism and the forearc region have been partially subducted beneath the Southern Uplands accretionary prism of the northern margin of Iapetus during end Caledonian collision. They base this view on the seismic velocity structure of continental basement beneath the southern part of the Southern Uplands which shows similarities with that beneath northern England but contrasts with the Midland Valley basement (Bamford, 1979). However the seismic data is far from clear and the presence of Midland Valley type basement beneath at least the northern part of the Southern Uplands (Upton et al. 1983; Hall et al. 1983) reduces the scope for northward subduction of the Lakesman accretionary prism. The Leggett et al. (1983) hypothesis minimises the need for strike-slip motion along the Iapetus Suture but raises the problem of subducting a wide zone (>100 kms) of buoyant crustal material. An alternative explanation, favoured here, would be to remove much of the Lakesman forearc and accretionary prism by major strike-slip along the suture.

The Lakesman rocks contain fossils belonging to a quite different faunal province from those further north (the Gondwanaland Province — see Dean, this volume). An important connection between these terranes, however, is again the presence of end Silurian — early Devonian granites in the Lakesman Terrane (Halliday et al. 1979). The Iapetus Suture in the British Caledonides appears to have been sealed by this time.

Monian Terrane

This is one of the smallest and most enigmatic terranes in the British Caledonides. It comprises the Monian rocks of Anglesey and Llŷn in NW Wales and the Cullenstown Formation in SE Eire. The terrane includes an extraordinarily wide variety of lithologies which include mudstones, turbidites, basic and acid volcanics, melange, granite, blueschists, greenschists and gneisses, all juxtaposed by major faults and mylonite zones (Gibbons 1983). They have been interpreted as exotic fragments produced by late Precambrian orogenic events along the active margin of an 'Avalonian' plate and subsequently moved by strike-slip faults during the late Precambrian or early Cambrian to their present position relative to the late Precambrian basement further SE (Gibbons 1984). After this eventual docking with the Southern Britain Terrane (see below), the Monian Terrain was affected by fault movements throughout the early Palaeozoic, particularly during the collisional end-Silurian event. Several deep crustal lineaments cutting the Monian Terrane, such as the Berw Fault in SE Anglesey, continued to influence sedimentation and tectonics long after the final disappearance of Iapetus. Both Devonian and Carboniferous sediments are locally strongly deformed where they have become involved in fault reactivation.

The contact between the Monian and Lakesman Terranes is obscured by younger rocks. The Monian terrane is overlain by Devonian red beds but unlike the terranes to the NW it was not intruded by end Silurian — early Devonian granites.

Southern Britain Terrane

The Precambrian rocks of southern and central Britain are very dissimilar to those exposed in the Monian Terrane. They consist predominantly of late Precambrian calc-alkaline plutonic and volcanic rocks and sediments, all of which were covered by a Lower to Middle Cambrian marine transgression (Thorpe 1979; Gibbons 1984). The thick succession of Welsh Basin Lower Palaeozoic sediments forms a cover to this late Precambrian basement. As shown by Bevins et al. (1984) the volcanic rocks erupted into the Welsh Basin are quite distinct from the calc-alkaline eruptives in the Lakesman Terrane. Whereas the early (Tremadoc) eruptions in the Welsh area are of calc-alkaline character, the later ones were not.

The Precambrian basement, which has been interpreted as an ancient volcanic arc (e.g. Thorpe 1979), is sometimes called 'Avalonian' because of direct correlation between these rocks and those in the Avalon Peninsula of Newfoundland (King 1980). This terrane appears to represent a small part of an Avalonian crustal plate now fragmented and widely dispersed through Canada, the eastern USA, and Europe (Piqué 1981). Although the southern part of this terrane virtually escaped the end-Silurian event, it was strongly affected by late Carboniferous (Hercynian) deformation.

The contact between the Monian and Southern Britain Terranes is either a major, steep, late fault (e.g. the Dinorwic Fault which bounds Anglesey to the SE) or is covered by Lower Palaeozoic sediments (e.g. the Arenig of the Llŷn Peninsula). The ancient Rosslare

Gneisses of SE Wexford lie SE of the 'Monian' Cullenstown Formation and are separated from it by a wide mylonite zone (Max 1975; Barber, Max and Brück 1981). The gneisses are cut by Cambrian (535 ± 6 Ma) and early Silurian (436 ± 6 Ma) granites (Max et al. 1979). They remain of uncertain affinity and could belong to the Monian, Southern Britain or Northern Armorican (i.e. Pentevrian — see below) Terranes, or some other unit unexposed elsewhere in the British Caledonides.

Northern Armorican Terrane

The Channel Islands and perhaps the (probably allochthonous) Lizard Peninsula of south Cornwall form part of the Armorican Massif of NW France. Unlike the basement in mainland S Britain (Hampton and Taylor 1983), this terrane includes late Archaean gneisses which are exposed on Guernsey and Sark (Adams 1967; Gibbons and Power 1975; Roach 1977). A succession of late Proterozoic metasediments (Brioverian) represent a cover to this basement and both are extensively intruded or overlain by late Precambrian/early Cambrian (Cadomian) igneous rocks of probable calc-alkaline character (Thorpe 1979). Cadomian plutonic rocks were intruded both before and after the deformation of the Brioverian metasediments (Adams 1967; Helm 1984). The relative position of this terrain to that of the Southern Britain Terrain during the Lower Palaeozoic is uncertain.

Comparison with Appalachian Terranes

The suspect terrain analytical approach has recently been applied to Appalachian geology (Williams and Hatcher 1982, 1983). As in Britain, the Appalachian Orogen comprises a Laurentian cratonic foreland upon and against which were accreted a series of terranes during the early Palaeozoic. Whereas some of the western Atlantic terranes may be traced into Britain with a fair degree of confidence (e.g. Avalon Terrane = Southern Britain Terrane), most of them have no obvious direct correlatives in the eastern Atlantic. This is particularly true of the terranes in the central part of the Orogen (although see suggestions by Kennedy 1979). The most widely made correlations, apart from the Avalonian, are those between the Dalradian and the Fleur de Lys Supergroups (e.g. Church 1969, Phillips et al. 1969, Kennedy 1972, 1975, 1979) and between the Monian and various Appalachian units (see Gibbons 1983, p. 158).

Despite these problems of direct along-strike correlation between terranes — problems which are unsurprising given the discontinuous, lensoid nature of most known suspect terranes — there are striking broad similarities in the timing of events affecting both the British Caledonian and the Appalachian orogens. The development of the Appalachian system was punctuated by three major Palaeozoic events (see Hatcher, this volume): the Penobscotian-Taconic (= Grampian), the Acadian (= 'end-Caledonian') and the Alleghanian (= Hercynian). The timing of these events was not exactly synchronous in the Appalachian and Caledonian sections of the orogen. There appears to have been some diachroneity from early to middle Ordovician in the Grampian-Taconic event, and from late Silurian to Devonian in the end-Caledonian-Acadian event. The Grampian-Taconic event involved the generation of regional metamorphic belts, and recumbent fold nappes and thrusts which verge NW and also (in the Grampian Terrane) SE. This important event is attributed to oceanward oceanic plate subduction perhaps followed by the collision of an arc system with the Laurentian foreland (Williams and Hatcher 1982; Mitchell 1984).

By comparison, the end Caledonian/Acadian event was relatively mild, with mostly upright structures and lower metamorphic grades. The difference between these two tectonic styles has been compared to the difference one might predict between 'head-on' (Taconic/Grampian) and oblique (Acadian/end-Caledonian) collisional events (Williams and Hatcher 1982). Following the disappearance of Iapetus in the British Caledonides and the NE Appalachians, a swarm of Devonian ('Newer') granites intruded the newly juxtaposed and thickened collage of Caledonian and Appalachian terranes (Strong 1980; Fyffe et al. 1981; Wones 1980; Thirlwall 1981).

Comparison with the Scandinavian Caledonides

Since Tuzo Wilson's (1966) Caledonian Ocean concept, it has been generally assumed that Southern Britain and Scandinavia formed part of the southeastern continental margin of Iapetus (e.g. Dewey 1969, Harland and Gayer 1972, Anderton et al. 1979). Adherence to this model makes it surprising that their Caledonian histories should be so different.

Scandinavian Caledonide tectonics are dominated by thrusting. Hossack and Cooper (in press) have divided the belt into four allochthonous, thrusted terranes. Each of these terranes is represented by a group of thrust sheets or nappes which have been thrust southeastwards by up to 500 km across the Baltic Craton. The two external, least travelled terranes are the Baltic Cover Thrust Sheets and the Baltic Crystalline Thrust Sheets, both of which were derived from within the Baltic Craton. The Oceanic Thrust Sheets Terrane lies to the northwest of and above the Baltic Crystalline Thrust Sheets and was derived from a broadly oceanic environment on the southeastern margins of Iapetus. The highest and furthest travelled terrane is the Exotic Nappes Terrane which is thought to have been derived either from microcontinents within Iapetus or from the Laurentian margin of the ocean, following final ocean closure and continental collision.

The boundaries between terranes are essentially low angle thrusts with strike slip movements being relatively unimportant and restricted to postcollisional faults. There are also differences in the timing of deformation events within the British and Scandinavian Caledonides. In Northern Scandinavia a major orogenic event, the Finnmarkian, occurred during the Mid Cambrian – Early Ordovician with the development of polyphase deformation, metamorphism and plutonism. Further south in Scandinavia the effects of the Finnmarkian event changed dramatically to ophiolite obduction southeastwards across the Baltic Shelf during Early-Mid Ordovician (initiation of the Oceanic thrust sheets). The Taconic-Grampian events of the Appalachians and North Britain are broadly coeval with the Finnmarkian but with contrasting polarity and developed on opposite margins of Iapetus. In southern Britain, the early signs of tectonism at this time are the activation of basin-controlling faults and volcanism, both attributed to the development of southeastward Iapetus subduction (Kokelaar et al. 1984).

The main collisional event in the Scandinavian Caledonides, the Scandian, occurred in Mid Silurian and resulted in the thrust telescoping of the Baltic Margin and the emplacement of the exotic nappes from the opposite margin of Iapetus. Thus the collision occured significantly earlier in Scandinavia than in Britain or the Appalachians. This discrepency could be accounted for by diachronous closure resulting from oblique convergent plate motions (e.g. Phillips et al. 1976) or, taken together with the other features of contrasting tectonic development, could indicate the presence of a plate boundary separating Britain from Scandinavia throughout the early Palaeozoic. Such a plate boundary has been suggested on the grounds of contrasting palaeoenvironments across the Tornquist Line (Spjeldnaes 1961, Cocks and Fortey 1982) and is indicated by the Caledonian radiometric dates of Palaeozoic rocks from North Sea drill holes (Ziegler 1978) and the subsurface geology in SW Poland (Ziegler 1981). It is further suggested by the discrepancy of palaeomagnetic poles throughout the Lower Palaeozoic (Piper 1983).

Independent movements of the Baltic Plate would result in the thrust terranes of the Scandinavian Caledonides being totally separate from those of the rest of the Caldeonide-Appalachian orogen, where thrusting on the southern margin of Iapetus is relatively insignificant (e.g. the 43 km of shortening across the Welsh Basin suggested by Coward and Siddans 1979). Recently, however, Leake et al. 1983, have identified a major thrust system in Connemara and suggested a SE vergence similar to that seen in the Scandinavian Caledonides.

Discussion

Potentially the most useful geological technique in the identification of suspect or exotic terranes is undoubtedly that of palaeomagnetism. Along the west coast of N. America this technique has successfully identified a highly complex mosaic of suspect terranes produced by prolonged oblique plate subduction and collision. However, the resolution of relative displacements using palaeomagnetism is, at best, several hundred kilometres, and E-W movements are undetectable. Unlike the western N. American plate boundary, the margins of the Iapetus Ocean in the British Area seem to have run in a NE-SW to E-W direction (e.g. Piper 1979; Smith et al. 1983). We would argue, therefore, that given the obvious differences in Lower Palaeozoic geology across most of the postulated terrane boundaries, the palaeomagnetic data cannot be used to support a lack of significant (i.e. at least several hundred kilometres) strike-slip movement.

Much controversy continues to centre on the amount of strike-slip movement which has taken place along the Great Glen Fault. The enormous (c. 2000 km) displacement postulated by Van der Voo and Scotese (1981) contradicts the more generally held view that movements along the fault have been restricted to <200 km (e.g. Kennedy 1946; Winchester 1974; Storetvedt 1974; Parnell 1982; Turnell and Briden 1982; Smith and Watson 1983). Given this interest, it is perhaps ironic that of all the steep terrane boundaries cutting the British Isles, the Great Glen Fault may have undergone the least amount of movement. Whereas geological features may be correlated across the Great Glen Fault (Smith and Watson 1983), pre-Devonian correlations across, for example, the Southern Uplands Fault, the Highland Boundary Fault, or the 'Iapetus Suture', are far more difficult.

The main conclusion of this paper is that the segmentation of the British Caledonides into a series of sub-parallel, NE-SW oriented linear belts, most of which are <120 km wide and each of which exhibit their own distinct pre-Devonian geological history, provides strong evidence in support of suspect terranes existing in this part of the orogen. To adopt the logic of a suspect terrane analyst, there is no strong evidence against major transcurrent fault movements having juxtaposed those British Caledonian Terranes bounded by steep faults. This process of orogenic terrane accretion began at least as early as the latest Precambrian (c.f. Monian and Southern Britain Terrains), with other major phases of activity occurring in the early Ordovician (Grampian) and the late Silurian (end-Caledonian) (Fig. 2). A particularly important phase of transcurrent movement is deduced to have taken place during the final disappearance of the Iapetus Ocean in the late Silurian. After this time, the terrane geometry of the British Caledonides remained essentially the same throughout the remaining Phanerozoic.

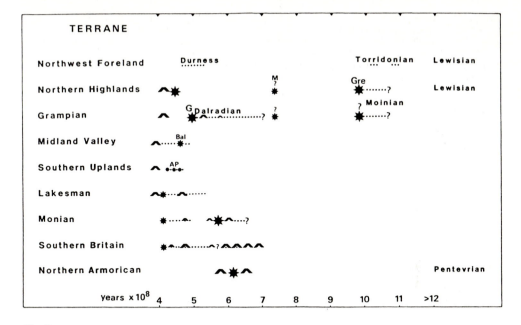

Fig. 2
Timing of events in British Caledonian Terranes.
Dotted lines = sedimentation; ∧ = igneous activity; stars = orogenic deformation and metamorphism, size of symbols represents approximate relative magnitudes; M = Morarian; Gre = Grenvillian; G = Grampian; Bal = Ballantrae ophiolite obduction; AP = Growth of accretionary prism (with attendant sedimentation, deformation and low grade metamorphism). The age ascribed to the Torridonian sediments is after Stewart et al. (1983).

Acknowledgments

We thank Richard Bevins, Dave Hughes, Graeme Taylor, Helen Turnell, and Graham Williams for their comments on this manuscript, and Paula Westall for word processing.

References

Adams, C.J.D. (1976): Geochronology of the Channel Islands and adjacent French mainland. J. geol. Soc. Lond., 132, 233–250.
Anderton, R., Bridges, P.H., Leeder, M.R., and Sellwood, B.W. (1979): A dynamic stratigraphy of the British Isles. George Allen & Unwin. 301 pp.
Atherton, M.P. (1977): The metamorphism of the Dalradian rocks of Scotland. Scott. J. Geol., 13, 331–70.
Barber, A.J., Max, M.D., and Brück, P.M. (1981): Field meeting in Anglesey and southeastern Ireland, 4–12 June 1977. Proc. Geol. Ass., 92, 269–92.
Bowes, D.R., Hopgood, A.M. (1975): Structure of the gneiss complex of Inishtrahull, C. Donegal. Proc. R. Ir. Acad., 75B, 369–390.
Bevins, R.E., Kokelaar, B.P., Dunkley, P.M. (1984): Petrology and geochemistry of early Ordovician igneous rocks in Wales: A volcanic arc to marginal basin transition. Proc. Geol. Ass., 95 (in press).
Brewer, M.S., Brook, M., and Powell, D. (1979): Dating of the tectono-metamorphic history of the southwestern Moine, Scotland. In: The Caledonides of the British Isles – reviewed, Harris, A.L., Holland, C.H., Leake, B.E. (eds.), Geol. Soc. Lond. Spec. Pub., 8, 129–137.

Briden, J.C., Morris, W.A., Piper, J.D.A. (1973): Palaeomagnetic studies in the British Caledonides — VI. Regional and global implications. Geophys. J.R. astr. Soc., 34, 107–134.

Brook, M., Powell, D., Brewer, M.S. (1977): Grenville events in Moine rocks of the Northern Highlands, Scotland. J. geol. Soc. Lond., 133, 489–496.

Brück, P.M., Colthurst, J.R.J., Feely, M., Gardiner, P.R.R., Penney, S.R., Reeves, T.J., Shannon, P.M., Smith, D.G., Vanguestaine, M. (1979): South-east Ireland: Lower Palaeozoic stratigraphy and depositional history. In: The Caledonides of the British Isles – reviewed, Harris, A.L., Holland, C.H. Leake, B.E. (eds.), Geol. Soc. Lond. Spec. Pub., 8, 533–544.

Cocks, L.R.M., Holland, C.H., Rickards, R.B., Strachan, I. (1971): A correlation of Silurian rocks in the British Isles. J. geol. Soc. Lond., 127, 103–136.

Cocks, L.R.M., Fortey, R.A. (1982): Faunal evidence for oceanic separations in the Palaeozonic of Britain. J. geol. Soc. Lond., 139, 467–480.

Coney, P.J., Jones, D.L., Monger, J.W.N. (1980): Cordilleran suspect terrains. Nature, 288, 329–33.

Coward, M.P. (1983): The thrust and shear zones of the Moine thrust zone and the NW Scottish Caledonides. J. geol. Soc. Lond., 140, 795–812.

Coward, M.J., Siddans, A.W.B. (1979): The tectonic evolution of the Welsh Caledonides. In: The Caledonides of the British Isles – review, Harris, A.L., Holland, C.H., Leake, B.E. (eds.), Geol. Soc. Lond. Spec. Pub., 8, 187–198.

Curry, G.B., Ingham, J.K., Bluck, B.J., Williams, A. (1982): The significance of a reliable Ordovician age for some Highland Border rocks in Central Scotland. J. geol. Soc. Lond., 139, 451–4.

Dewey, J.F. (1969): Evolution of the Appalachian/Caledonian Orogen. Nature, London, 222, 124–9.

Dewey, J.F. (1982): Plate tectonics and the evolution of the British Isles. J. geol. Soc. Lond., 139, 371–414.

Elliott, D., Johnson, M.R.W. (1980): Structural evolution in the northern part of the Moine Thrust Zone. Trans. R. Soc. Edinburgh, 71, 69–96.

Flinn, D., May, F., Roberts, J.L., Treagus, J.E. (1972): A revision of the stratigraphic succession of the East Mainland of Shetland. Scott. J. Geol., 8, 335–43.

Flinn, D., Frank, P.L., Brook, M., Pringle, I.R. (1979): Basement-cover relations in Shetland. In: The Caledonides of the British Isles – reviewed, Harris, A.L., Holland, C.H., Leake, B.E. (eds.), Geol. Soc. Lond. Spec. Pub., 8, 109–116.

Fyffe, L.R., Pajari, G.E., Cherry, M.E. (1981): The Acadian plutonic rocks of New Brunswick. Maritime sediments and Atlantic Geology, 17, 23–26.

Garson, M.S., Plant, J. (1972): Possible dextral movements on the Great Glen and Minch Faults, Scotland. Nature, Phys. Sci., 240, 31–32.

Gibbons, W. (1983): Stratigraphy, subduction and strike-slip faulting in the Mona Complex of North Wales – a review. Proc. Geol. Ass., 94, 147–163.

Gibbons, W. (1984): The Precambrian basement to England and Wales. Proc. Geol. Ass., 95, in press.

Gibbons, W., Power, G.M. (1975): The structure and age of the gneisses of Sark, Channel Islands. Proc. Ussher Soc., 3, 244–251.

Hall, J., Powell, D.W., Warner, M.R., El-Isa, Z.H.M., Adesanym, O. and Bluck, B.J. (1983): Seismological evidence for shallow crystalline basement in the Southern Uplands of Scotland. Nature, London, 305, 418–20.

Halliday, A.N., Aftalion, M., van Breeman, O., Jocelyn, J. (1979): Petrogenetic significance of Rb-Sr and U-Pb isotopic systems in the 400 Ma old British Isles granitoids and their hosts. In: The Caledonides of the British Isles – reviewed, Harris, A.L., Holland, C.H., Leake, B.E. (eds.), Geol. Soc. Lond. Spec. Pub., 8, 653–662.

Hampton, C.M., Taylor, P.N. (1983): The age and nature of the basement of southern Britain: evidence from Sr and Pb isotopes in granites. J. geol. Soc. Lond., 140, 499–509.

Harland, W.B., Gayer, R.A. (1972): The Arctic Caledonides and earlier oceans. Geol. Mag., 109, 281–314.

Harris, A.L., Pitcher, W.S. (1975): The Dalradian supergroup. In: A correlation of Precambrian rocks in the British Isles. Spec. Rep. geol. Soc. Lond., 6, 52–75.

Helm, D.G. (1984): The tectonic evolution of Jersey, Channel Islands. Proc. Geol. Ass., 95, 1–16.

Henderson, W.G., Robertson, A.H.F. (1982): The Highland Border rocks and their relation to marginal basin development in the Scottish Caledonides. J. geol. Soc. Lond., 139, 433–50.

Higgins, A.C. (1967): The age of the Durine Member of the Durness Limestone Formation at Durness. Scott. J. Geol., 3, 382–8.

Hossack, J.R., Cooper, M. (in press): Collisional tectonics in the Scandinavian Caledonides. J. geol. Soc. Lond.

Ikin, N.P. (1983): Petrochemistry and tectonic significance of the Highland Border Suite mafic rocks. J. geol. Soc. Lond., 140, 267–279.

Kennedy, W.Q. (1946): The Great Glen Fault. Q. J. geol. Soc. Lond., 102, 41–72.
Kennedy, M.J. (1975): The Fleur de Lys Supergroup: stratigraphic comparison of Moine and Dalradian equivalents in Newfoundland with the British Caledonides. J. geol. Soc. Lond., 131, 305–10.
Kennedy, M.J. (1979): The continuation of the Canadian Appalachians into the Caledonides of Britain and Ireland. In: The Caledonides of the British Isles – reviewed, Harris, A.L., Holland, C.H., Leake, B.E. (eds.), Geol. Soc. Lond. Spec. Pub. 8, 33–64.
King, A.F. (1980): The birth of the Caledonides: Late Precambrian rocks of the Avalon Peninsula, Newfoundland, and their correlatives in the Appalachian-Orogen. In: The Caledonides in the U.S.A., Wones, D.R. (ed.), Virginia Polytechnic Institute and State University Memoir 2, 3–8.
Kokelaar, B.P., Howells, M.F., Bevins, R.E., Roach, R.A., Dunkley, P.M. (1984): The Ordovician Marginal Basin of Wales. In: Marginal Basin Geology: volcanic and associated sedimentary and tectonic processes in modern and ancient marginal basins. Geol. Soc. Lond. Spec. Pub. (in press).
Leake, B.E., Tanner, P.W., Singh, D. (1983): Major southward thrusting of the Dalradian rocks of Connemara, Western Ireland. Nature, 305, 210–213.
Leggett, J.K., McKerrow, W.S., Eales, M.H. (1979): The Southern Uplands of Scotland: a Lower Palaeozoic accretionary prism. J. geol. Soc. Lond., 136, 755–70.
Leggett, J.K., McKerrow, W.S., Soper, N.J. (1983): A model for the crustal evolution of Southern Scotland. Tectonics, 2, 187–210.
Macintyre, R.M., Van Breemen, O., Bowes, D.R., Hopgood, A.M. (1975): Isotopic study of the gneiss complex, Inishtrahull, Co. Donegal. Scient. Proc. R. Dub. Soc. Ser. A, 5, 301–309.
Max, M.D. (1975): Precambrian rocks of SE Ireland. In: A correlation of the Precambrian rocks in the British Isles, Harris, A.L., Shackleton, R.M., Watson, J., Downie, C., Harland, W.B., Moorbath, S., (eds.), Geol. Soc. Lond. Spec. Rep., 6, 97–101.
Mitchell, A.H.G. (1984): The British Caledonides: interpretations from Cenozoic analogues. Geol. Mag., 121, 35–46.
Molyneux, S.G. (1979): New evidence for the age of the Manx Group, Isle of Man. In: The Caledonides of the British Isles – reviewed, Harris, A.L., Holland, C.H., Leake, B.E. (eds.), Geol. Soc. Lond. Spec. Pub., 8, 415–422.
Moseley, F. (1978): The geology of the Lake District. Yorks. geol. Soc. p. 284.
Parnell, J.T. (1982): Comment on Palaeomagnetic evidence for a large (2000 km) sinistral offset along the Great Glen Fault during Carboniferous time. Geology, 10, 605.
Parsons, I., McKirdy, A.P. (1983): Inter-relationships of igneous activity and thrusting in Assynt: excavations at Loch Borrolan. Scott. J. Geol., 19, 59–66.
Phillips, W.E.A. (1981): The Pre-Caledonian basement. The Orthotectonic Caledonides. In: A geology of Ireland, Holland, C.H. (ed.), 7–40.
Phillips, W.E.A., Stillman, C.J., Murphy, T. (1976): A Caledonian plate tectonic model. J. geol. Soc. Lond., 132, 579–609.
Piasecki, M.A.J. (1980): New light on the Moine rocks of the Central Highlands of Scotland. J. geol. Soc. Lond., 137, 41–59.
Piasecki, M.A.J., Van Breeman, O., (1979): A Morarian age for the 'younger Moines' of central and western Scotland. Nature, London, 278, 734–6.
Piasecki, M.A.J., Van Breemon, O., Wright, A.E. (1981): Late Precambrian geology of Scotland, England and Wales. Canad. Soc. Petrol. Memoir 7, 57–94.
Piper, J.D.A. (1979): Aspects of Caledonian palaeomagnetism and their tectonic implications. Earth planet. Sci. Lett., 44, 176–192.
Piper, J.D.A. (1983): Palaeomagnetism in the Caledonian-Appalachian Orogen: A Review. In: The Caledonide Orogen, Gee, D.G., Sturt, B.A. (eds.), Wiley (Lond.) (in press).
Piqué, A. (1981): Northwestern Africa and the Avalonian plate: relation during late Precambrian and late Palaeozoic time. Geology, 9, 319–22.
Powell, D., Brook, M., Baird, A.W. (1983): Structural dating of a Precambrian pegmatite in Moine rocks of northern Scotland and its bearing on the status of the 'Morarian Orogeny'. J. geol. Soc. Lond., 140, 813–824.
Rathbone, P.A., Harris, A.L. (1979): Basement-cover relationships at Lewisian inliers in the Moine rocks. In: The Caledonides of the British Isles – reviewed, Harris, A.L., Holland, C.H., Leake, B.E. (eds.), Geol. Soc. Lond. Spec. Pub., 8, 101–108.
Roach, R.A. (1977): A review of the Precambrian rocks of the British Variscides and their relationships with the Precambrian of N.W. France. In: La chaine varisque d'Europe moyenne et occidentale. Coll. inter., C.N.R.S. Rennes, 243, 61–79.
Ryan, P.D., Sawal, V.K., Rowlands, A.S. (1983): Ophiolitic mélange separates ortho- and paratectonic Caledonides in western Ireland. Nature, 302, 50–52.

Smith, A.G., Hurley, A.M., Briden, J.C. (1981): Phanerozoic palaeocontinent world maps. Cambridge University Press. p. 102.
Smith, D.I., Watson, J. (1983): Scale and timing of movements on the Great Glen Fault, Scotland. Geology, 11, 523–526.
Smith, R.L., Stearn, J.E.F., Piper, J.D.A. (1983): Palaeomagnetic studies of the Torridonian sediments, NW Scotland. Scott. J. Geol., 19, 29–46.
Spjeldnaes, N. (1961): Ordovician climatic zones. Norsk. Geol. Tidsskr., 41, 45–77.
Storetvedt, K.M. (1973): A possible large-scale sinistral displacement along the Great Glen Fault in Scotland. Geol. Mag., 111, 23–30.
Swett, K. (1981): Cambro-Ordovician strata in Nyfriesland, Spitzbergen and their palaeotectonic significance. Geol. Mag., 118, 225–250.
Thirlwall, M.F. (1981): Implication for Caledonian plate tectonic models of chemical data from volcanic rocks of the British Old Red Sandstone. J. geol. Soc. Lond., 138, 123–38.
Thorpe, R.S. (1979): Late Precambrian igneous activity in S. Britain. In: The Caledonides of the British Isles – reviewed, Harris, A.L., Holland, C.H., Leake, B.E. (eds.), Geol. Soc. Lond. Spec. Pub., 8, 579–84.
Turnell, H.B., Briden, J.C. (1983): Palaeomagnetism of NW Scotland Syenites in relation to local and regional tectonics. Geophys. J. R. astr. Soc., 75, 217–234.
Upton, B.J.G., Aspen, P., Chapman, N.A. (1983): The upper mantle and deep crust beneath the British Isles: evidence from inclusions in volcanic rocks. J. geol. Soc. Lond., 140, 105–122.
Van Breeman, O., Pidgeon, R.T., Johnson, M.R.W. (1974): Precambrian and Palaeozoic pegmatites in the Moines of Northern Scotland. J. geol. Soc. Lond., 130, 493–507.
Van der Voo, R., Scotese, C. (1981): Palaeomagnetic evidence for a large (2000 km) sinistral offset along the Great Glen Fault during Carboniferous time. Geology, 9, 583–589.
Williams, H., Hatcher, R.D. (1982): Suspect terranes and accretionary history of the Appalachian orogen, Geology, 10, 530–536.
Wilson, J.T. (1966): Did the Atlantic close and then re-open? Nature, 211, 676–681.
Winchester, J.A. (1974): The zonal pattern of regional metamorphism in the Scottish Caledonides. J. geol. Soc. Lond., 130, 509–524.
Wones, D.R. (1980): A comparison between granitic plutons of New England, USA and the Sierra Nevada batholith, California. In: The Caledonides in the USA, Virginia Polytechnic Institute and State University Memoir 2, Wones, D.R. (ed.), 123–130.
Ziegler, P.A. (1978): North-western Europe: tectonics and basin development. In Key-notes of the MEGS-II, Amsterdam, van Loon, A.J. (ed.), Geol. en Mijnb., 57, 589–626.

Revised manuscript received 25. Febr. 1984

Relationships of Cambrian-Ordovician Faunas in the Caledonide-Appalachian Region, with Particular Reference to Trilobites

W. T. Dean
Department of Geology, University College, Cardiff, U.K.

Key Words
Cambrian
Ordovician
Iapetus
Trilobites
Faunal provinces

Abstract

Relationships of Cambrian and Ordovician faunas in the Caledonide-Appalachian region are discussed with reference to three principal continental regions: Laurentia, Gondwanaland and Baltica. Cambrian faunas and rocks on either side of Iapetus are generally distinct, but the *Fallotaspis* Zone (early Lower Cambrian) is widespread, being found in Morocco, western North America and Siberia. Middle Cambrian trilobites from a single locality in the central Appalachians comprise two genera of wide geographic distribution and do not necessarily provide evidence of 'mixed faunas'. Middle and Upper Cambrian faunas in Baltica, Wales and the eastern Appalachians have much in common and often occur in dark shales. Owing to the influence of facies they show few links with Middle Cambrian faunas in the Mediterranean region, where Upper Cambrian trilobites are absent or exhibit Asiatic affinities.

Tremadoc trilobite faunas on either side of Iapetus, though generally different, can be correlated by means of the cosmopolitan dendroid graptolite *Rhabdinopora flabelliformis*. Gondwanaland Arenig trilobites occur mostly in clastic rocks and include groups, such as trinucleids and calymenids, not known at this level in Laurentia or Baltica. Laurentian carbonate faunas comprise particularly the bathyurid trilobites, including *Petigurus* in northwest Scotland. In Wales Gondwanaland faunas are associated with elements found mostly around the margins of Laurentia; this suggests either that some trilobites and brachiopods were capable of unusually extensive migrations, or that parts of Iapetus may have been narrower than has been supposed. Increasing proximity of western European part of Gondwanaland and Laurentia in the Caradoc is indicated by both trilobite and brachiopod faunas. The distance between Baltica and Gondwanaland also decreased at this time. Subsequent almost cosmopolitan distribution of trilobites in the Ashgill was affected by extensive glaciation, centred on North Africa, in the Hirnantian; as a result numerous trilobite families with a previously long history became extinct, and the importance of the class was considerably diminished from the Silurian onwards.

Introduction

In recent years it has become routine practice to show palaeogeographic reconstructions of England and Wales during Cambrian and Ordovician times in relation to an Iapetus Ocean (Harland & Gayer, 1972) that probably widened during at least part of the Cambrian and closed progressively during much of the Ordovician. The term Iapetus replaced the 'Proto-Atlantic' of Wilson (1966), whose views were much influenced by the presence of 'Atlantic' or 'European' faunas in eastern North America, and of North American or 'Pacific' faunas in northwest Scotland. The existence of such faunas has long been appreciated and was invoked by Wilson when postulating the closure of his 'Proto-Atlantic', but it is only since the general acceptance of plate tectonic theory that their distribution has become intelligible in palaeogeographic terms.

Trilobites form the basis of most of this discussion but other fossil groups are noted where appropriate. Whether or not one believes in the concept of faunal provinces in the fossil record, the faunal assemblages discussed here were distributed, often with remarkable uniformity, around the periphery of enormous continental masses (Cowie, 1971) such as Laurentia and Gondwanaland in a manner unparalleled at the present day. The nature and composition of the faunas were influenced both by the palaeoclimate, particularly in terms of latitude, and by the sedimentary environment in relation to the adjacent continental shoreline. Both these factors have in turn to be considered in relation to the palaeomagnetic evidence, and sometimes there are differences of interpretation that have still to be resolved. Space does not permit a detailed consideration of all these topics but this paper summarises the main faunal distributions in the Caledonide region, and occasionally elsewhere, and reviews some of the unsolved problems.

Comments on Palaeogeographic Regions and 'Faunal Realms'

Laurentia

Comprises the Precambrian shield of North America with the addition of Greenland, Spitsbergen, northwestern Norway and northwestern Scotland, particularly that part delimited by the Moine Thrust but including also the area northwest of the Highland Boundary Fault and its southwesterly extension into Ireland. The region contains the so-called 'Pacific' faunas (a term not used in this account) of Cambrian age, and for most of the Ordovician coincides with the distribution of Whittington & Hughes's (1972) Bathyurid (trilobite) Province. During the Cambrian the margins of Laurentia were the site of broad carbonate banks that separated a landward region of light-coloured, shallow-water, terrigenous rocks (inner detrital belt) from a seaward region (outer detrital belt) with dark shales, mudstones and thin-bedded limestones. These terms were introduced and summarised by Palmer (1969, 1979) who showed that the trilobite faunas of the inner detrital belt and the inner margin of the carbonate bank were of restricted type, distinct from those of the outer bank and outer detrital belt, which had access to the 'open ocean' and were often very widely distributed geographically. Carbonate sedimentation on and around Laurentia continued during most of the Ordovician and may follow the Cambrian pattern, but the terms inner and outer detrital belts are less generally applicable.

Gondwanaland

Although proposed originally for an Upper Palaeozoic 'supercontinent' comprising Africa, South America, India, Australia and Antarctia, Gondwanaland probably existed from as long ago as the Proterozoic until its fragmentation in the Mesozoic (for review of literature see Windley, 1977, pp. 141, 147, 203). The term is certainly useful when discussing Lower Palaeozoic faunas, though with its geographic boundaries slightly modified from the above so as to encompass at least, inter al., England (including the Lake District), Wales, much of present day Europe, southern Turkey, the Arab Platform and southwestern China. In biogeographic terms Gondwanaland coincides broadly with the Calymenid-Trinucleid Province of Whittington (1966), later the *Selenopeltis* Province of Whittington and Hughes (1972), a unit containing faunas, variously termed Bohemian, Mediterranean or Tethyan, that inhabited a cool-water region of essentially clastic sedimentation, particularly of dark muds and silts with subsidiary sands and, in the Welsh area and the Lake District, volcanic rocks. Whittington and Hughes's (1972) use of a Proto-Tethys separating southern Europe from the main mass of Gondwanaland seems to be unsubstantiated and, as on previous occasions (Burrett, 1973; Dean, 1976), is not included here. World maps for the Cambrian and Ordovician published by Smith et al. (1981) show England, Wales, much of Europe, including Iberia, and the northwest tip of Africa far removed from the main mass of Gondwanaland, the area of separation coinciding with, though considerably larger than, Whittington and Hughes's Proto-Tethya. This interpretation, too, is not incorporated here because of the stratigraphic and palaeontological evidence linking England, Wales and adjacent Europe with the Mediterranean region during the Cambrian and Ordovician. Further evidence for the palaeogeographic continuity of Gondwanaland is provided by microfossils such as acritarchs (see review by Martin, 1982), a group important for correlation in areas where thick Cambrian and Ordovician clastics occur but contain no macrofossils.

Baltica

Coincides essentially with the Asaphid Province of Whittington and Hughes (1972), defined by them as comprising the Eastern European Platform of Størmer (1967), the Russian Platform, Ural Mountains, Pay Khoy and Novaya Zemlya. This usage differs from the Baltica of Ziegler and others (1977), who included also England, Wales and southeastern Ireland. At the present day Baltica is bounded in the west by the main line of Caledonian thrusting, and may be separated from western Europe (i.e. part of Gondwanaland in the present paper) by the structure known as Tornquist's Lineament or Line (Fig. 1), so named for the German geologist A. Tornquist, not to be confused with the Swedish palaeontologist S.K. Törnquist. The position of the southwestern boundary of Baltica has been variously interpreted (Størmer, 1967; Strömberg, 1981). Cocks and Fortey (1982, p. 467) postulated a 'Tornquist's Sea' separating Britain and Baltica (= Gondwanaland and Baltica of this paper) in the early Ordovician (Arenig). Their Baltica boundary does not coincide with Tornquist's Lineament (or with any of the other boundaries reviewed by Størmer, 1967), but runs subparallel to and southwest of it. The interpretation found more convenient here places Bornholm and southernmost Sweden with Baltica and not with Gondwanaland.

During the Ordovician, Baltica was the site of carbonate sedimentation (Jaanusson, 1972) that took place mostly in a temperate climatic zone, though some marginal deposits were of subtropical, Bahaman type. Within these environments lived the often thick-shelled asaphid trilobites, including *Asaphus* s.s., of Whittington and Hughes's Asaphid Province.

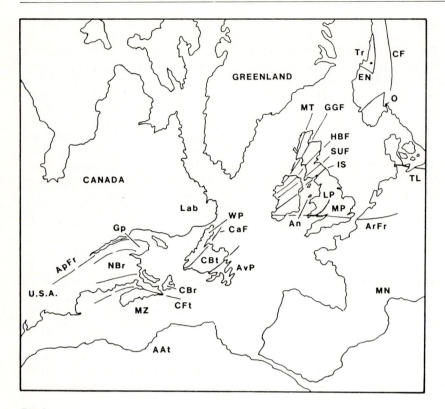

Fig. 1
Outline map, based on a closed Iapetus Ocean, showing selected geographic names and geological structures in western Europe, northwest Africa, Scandinavia and eastern North America.

AAt, Anti-Atlas Mts.; ApFr, Appalachian Front (= Logan's Line); An, Anglesey; ArFr, Armorican Front; AvP, Avalon Platform; CaF, Cabot Fault; CBr, Cape Breton Island; CBt, Central Mobile Belt of Newfoundland; CF, Caledonide Front; CFt, Chedabucto Fault; EN, exotic nappes of the Trondheim Nappe Complex; GGF, Great Glen Fault; Gp, Gaspé Peninsula; HBF, Highland Boundary Fault; IS, Iapetus Suture; Lab, Labrador; LP, Linley-Pontesford lineament; MN, Montagne Noire; MP, English Midland Platform; MT, Moine Thrust; MZ, Meguma Zone; NBr, New Brunswick; O, Oslo; SUF, Southern Upland Fault; Tr, Trondheim; TL, Tornquist's Line; WP, Western Platform of Newfoundland.

It should be noted, however, that asaphids s.l., far from being confined to the Asaphid Province were often abundant in clastic sediments deposited in cool waters around Gondwanaland, as well as occurring in shales and mudstones off the margins of Laurentia, that is to say in a position approximating to the outer detrital belt of the Cambrian.

Other Terms

The adjective 'Acado-Baltic' is commonly applied to Cambrian trilobite faunas in England, Wales, Scandinavia, the Baltic area and eastern Appalachians. It was named for Acadia (originally the French Acadie), which comprised the Canadian provinces of Nova Scotia (including Cape Breton Island) and part of neighbouring New Brunswick, together with the Baltic area in its broad sense, including most of Norway, Sweden and the Danish island of Bornholm. It takes in also a portion of the eastern Appalachians in U.S.A. that includes part of Massachussetts, Vermont and Pennsylvania. Atlantic Faunas and Atlantic Province, often used synonymously with Acado-Baltic, are considered inappropriate here, as is Pacific Faunas, for which the term Laurentian is preferred. The use of 'Acado-Baltic' implies a continuity between the eponymous regions that may not be justified, though some degree of marine connection existed during most of the Cambrian.

Cambrian

The Cambrian as used here excludes the Tremadoc Series, and its threefold subdivision into Lower, Middle and Upper broadly follows Scandinavian usage. Corresponding Series names, successively Comley, St. David's and Merioneth, introduced by Cowie and others (1972) are noted later; their use presents some problems and it is questionable whether they will be accepted internationally.

Lower Cambrian (Fig. 2)

The fullest known development of trilobite-bearing Lower Cambrian strata is in the Anti-Atlas of Morocco, where Hupé (1953) documented a succession of eight zones; the oldest (Zone I) was named for the olenellid *Fallotaspis tazemmourtensis*, but the genus was

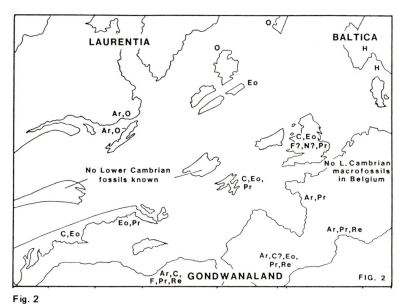

Fig. 2

Lower Cambrian: Ar, Archaeocyatha; C, *Callavia*; Eo, Eodiscina; F, *Fallotaspis*; H, *Holmia*; N, *Nevadella*; O, *Olenellus*; Pr, Protolenidae; Re, Redlichiidae.

Figs. 2–9. Generalised outline maps showing distribution of selected faunal elements in the Iapetus region. Owing to space limitations the width of Iapetus is not drawn to scale. Longitudinal position of structural units is arbitrary and takes little account of lateral movements that must have taken place but are inadequately known.

recorded from Zones I to IV, subsequently considered as subzones within a *Fallotaspis* Zone introduced by Fritz (1972) and based on the Moroccan succession. More recently Sdzuy (1978) described from rocks in the Anti-Atlas older than *Fallotaspis* Zone a new genus *Eofallotaspis* (possibly ancestral to *Fallotaspis*) that was underlain by still older, undescribed opisthoparian trilobites. No formal zone was designated for *Eofallotaspis* and the genus is unknown outside the Anti-Atlas. The *Fallotaspis* Zone is very widely distributed, being recorded from southeastern California (Nelson and Hupé, 1964), where *Fallotaspis* occurs with *Daguinaspis* (described from Zone III of Morocco by Hupé, 1953), and from both western and northwestern Canada, where *Fallotaspis* may be accompanied by *Parafallotaspis* Fritz (1972, p. 5). In England a single olenellid from the so-called '*Holmia* Sandstone', (Ab 3) or Lower Comley Sandstone of Shropshire was assigned questionably to *Fallotaspis* by Hupé (1953). The palaeogeographic distribution of these presumably benthic forms in both Gondwanaland (Anti-Atlas; Comley) and the cordilleran region of Laurentia is not yet satisfactorily explained. It is difficult to acommodate within reconstructions such as that by Smith, Hurley and Briden (1981, map 80) and may indicate that Iapetus was narrower in the Lower than in the Middle and Upper Cambrian.

The term '*Holmia-Callavia* fauna' has often been used to encompass both Baltica and Wales but the evidence for palaeogeographic continuity of these areas and their benthic faunas is inconclusive and there may have been some degree of separation. Scandinavian Lower Cambrian faunas (Ahlberg, 1981; Bergström, 1981), though incompletely known, include *Holmia* (an olenellid) and *Strenuaeva*, genera not yet described from England and Wales; conversely *Callavia*, long known from Shropshire, has not been recorded from Scandinavia, though *Comluella*, founded by Hupé (1953) on a species from the late Lower Cambrian of Comley, is found in Sweden (Bergström, 1981, p. 24). In the eastern Appalachians *Callavia* is found in the Avalon Platform (its type area), eastern Newfoundland, at the same horizon as at Comley (Hutchinson, 1962), as well as in the Boston area, Massachusetts (Palmer, 1971, p. 211). Coincident with the upper part of the stratigraphic range of *Callavia* there occurs the pagetiid *Dipharus attleborensis*, described first from Massachusetts and index species of an assemblage said by Bergström (1981) to be represented in Sweden. The eodiscid *Serrodiscus bellimarginatus*, also from Massachusetts, occurs in eastern Newfoundland and in the '*Eodiscus* Limestone' of Comley. Still further evidence of close affinity between the eastern Appalachians and Wales is provided by records of the conocoryphid *Pseudatops* from the Lower Cambrian of Salem, New York, the *Callavia* Zone of the Avalon Platform, the Llanberis Slates of northwest Wales, and the *Serrodiscus bellimarginatus* Limestone of Comley.

Introduction by Cowie and others (1972) of a Protolenid-Strenuellid Zone as topmost subdivision of the Lower Cambrian in England and Wales reflects on the one hand the relative abundance of these two trilobite groups in palaeogeographically related areas, and on the other the difficulty of choosing an appropriate zonal index. The Protolenidae, though in need of modern revision, provide evidence of biogeographic continuity between parts of Gondwanaland, and the family does not occur in Laurentian faunas. In northwest Wales Bassett and others (1976) assigned material from the Llŷn Peninsula to the Moroccan subgenus *Hamatolenus (Myopsolenus)*. *Protolenus* s.s., first described from shales in southern New Brunswick, has been widely reported from Gondwanaland, but some of the records need to be substantiated.

Lower Cambrian faunas on the Laurentian side of Iapetus were at one time considered to fall within a single '*Olenellus* Zone'. Work in western North America by Fritz (1972) showed that *Olenellus* occurs only in his *Bonnia-Olenellus* Zone, of topmost Lower

Cambrian age, approximately equivalent to the Protolenid-Strenuellid Zone and thus younger than the Olenellid Zone proposed by Cowie and others (1972) for the Welsh area.

In the British Isles *Olenellus* and other trilobites of Laurentian type have been found only in the 'Fucoid' Beds and overlying *Salterella* Grit of the Durness area, northwest of the Moine Thrust, but similar faunas were distributed around the whole margin of Laurentia (Cowie, 1971). The succeeding Durness Group (Cowie and others, 1972), mostly limestones and dolomites, exhibits both lithological and faunal affinities with parts of Laurentia, including east Greenland, Spitsbergen and, especially, western Newfoundland. Correlation of the eight formations constituting the Durness Group is imprecise but no unequivocal Middle or Upper Cambrian faunas are recorded and a significant disconformity must exist. The three topmost formations are correlated (Whittington in Williams and others, 1972) almost entirely with the Arenig Series and possibly the lowest Llanvirn Series.

In central Scotland, between the Highland Border Fault and the Great Glen Fault, lies the outcrop of the Dalradian metasediments, raised by Cowie and others (1972) to the status of a supergroup and extending southwest across Ireland. The enormous thickness (12 000 m) is divisible into lower, middle and upper, but fossils are extremely rare and correlation is tenuous. Trace fossils in part of the middle Dalradian resemble those from Lower Cambrian rocks at Durness that contain *Olenellus*. In the upper Dalradian the small trilobite *Pagetides* (Suborder Eodiscina) from the Leny Limestone indicates affinities with faunas in eastern USA (New York State) and Canada (Quebec) originally sited on the Laurentia side of Iapetus. The age of the Leny Limestone (Cowie and others, 1972, p. 17) in terms of the Welsh area is Protolenid-Strenuellid Zone. No similar faunas are known from the Caledonides, nor is there evidence of post-Lower Cambrian strata in the Dalradian Supergroup, but in view of the thickness above the Leny Limestone Cowie and others (1972) did not exclude the possibility of an age extending into the early Ordovician.

Middle and Upper Cambrian (Figs. 3, 4)

The Middle Cambrian of Europe and Scandinavia coincides at least approximately with the vertical range of the trilobite *Paradoxides* s.l., and the most satisfactory 'standard' succession is that of Sweden, where Westergård (1946) proposed three successive 'Stages' of *P. oelandicus, P. paradoxissimus,* and *P. forchhammeri;* these units have since been termed zones by Rushton (1966, p. 6) and, once again, stages by Martinsson (1974). In Wales and adjacent areas the Lower/Middle Cambrian boundary generally coincides with an unconformity or with arenaceous strata that lack trilobites. Detailed correlation, particularly intercontinentally, relies heavily on the more refined scale of nine zones, six of them based on agnostid trilobites, established by Westergård (1946) for the Swedish succession.

The St. David's Series, equivalent to Middle Cambrian as usually understood, was introduced by Cowie and others (1972) in an attempt to stabilise chronostratigraphic subdivision of the Cambrian in the Welsh area. The term is difficult to apply in its entirety as both lower and upper boundaries are ill defined in the type area of southwest Wales where the Caerfai Group, oldest of the local Cambrian rocks, rests unconformably on Precambrian (Pebidian) volcanic rocks. The Caerfai Group has not been satisfactorily dated by means of fossils and, from its stratigraphic position when compared with the succession in eastern Newfoundland, could well be early Middle Cambrian rather than Lower Cambrian as has been claimed.

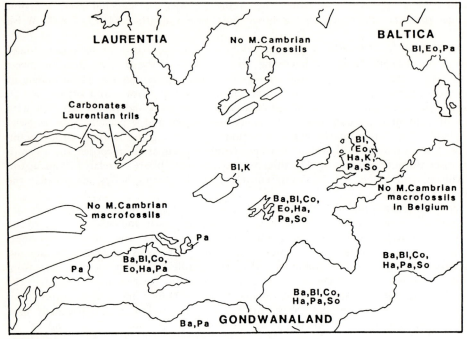

Fig. 3
Middle Cambrian: Ba, *Badulesia;* Bl, *Bailiella;* Co, *Conocoryphe;* Eo, *Eodiscina;* Ha, *Harttella;* K, *Kootenia;* Pa, *Paradoxides* s.l.; So, *Solenopleuropsis.*

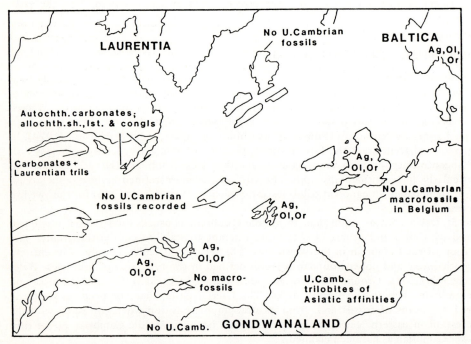

Fig. 4
Upper Cambrian: Ag, Agnostidae; Ol, Olenidae; Or, *Orusia.*

Dark, often anoxic sediments are widespread in the Middle Cambrian of Wales and, in more condensed form, of Scandinavia, a distribution broadly comparable with that in the Upper Cambrian and extending subparallel to the Caledonian Front. Merioneth Series was proposed by Cowie and others (1972) for the relatively thick Upper Cambrian clastics in North Wales where the rocks are probably too poorly fossiliferous or too cleaved to provide an internationally acceptable stratotype. Nevertheless it has proved possible to apply there with some success the detailed zonal scheme founded in Scandinavia (Henningsmoen, 1957) on thinner sequences of dark, anoxic mudstones and shales with occasional dark limestones (stinkstones) in which olenid trilobites occur, often in large numbers, with agnostids in a subsidiary role. Opinion on the mode of life of agnostids has shifted from considering them as 'presumably mud burrowers' (Wilson, 1957, p. 336) to an appreciation of their 'pelagic mode of life in the oceanic province' (Robison, 1972, p. 33), an assessment more in keeping with their widespread geographic distribution and increasing importance in international correlation. Opinions vary as to the mode of life of olenids. Henningsmoen (1957) distinguished three morphological groups, ranging from active to less active swimmers to planktonic forms. Bergström (1973) considered the olenids to be 'swimming animals' and Ross (1975) regarded them as pelagic; Fortey and Owens (1978) disagreed, however, arguing for a benthic mode of life with life habits 'closely tied to bottom or near-bottom conditions'. None of these cases is yet proven. The delicate, highly spinose exoskeleton of *Ctenopyge,* for example, might have been more at home near the surface of the sea; on the other hand it would have had no competitors in the low-energy, anoxic sea-floor environment, to which olenids apparently were adapted, that produced the Upper Cambrian black shales.

Palmer (1973) included Agnostidae, Eodiscidae, Olenidae, Conocoryphidae and Paradoxididae in a selection of 'trilobites characteristic of open ocean regions', with the tacit implication of a pelagic mode of life enabling those groups to traverse areas of deeper water that presumably would act as barriers to trilobites of benthic type, the latter termed by him 'characteristic of restricted regions'. Various Middle Cambrian agnostid and eodiscid species are common to Baltica, England and Wales, the Avalon Platform of Newfoundland and related parts of the eastern Appalachians, but evidence of mobility of *Paradoxides* species between these areas is less conclusive. Such well-known Welsh species as *P. hicksii* and *P. davidis* are abundant at some levels in the Manuels River Formation of eastern Newfoundland (Hutchinson, 1962; Bergström and Levi-Setti, 1978), in dark shales and mudstones, and dark-grey, lenticular limestones. Neither of these species is in Hayes and Howell's (1937) faunal lists for the Saint John area, southern New Brunswick, nor are they listed from most of the classic sections in Scandinavia, though Grönwall (1902) described *P. davidis* and *P. hicksii* from Bornholm. The Scandinavian Paradoxididae still await monographic revision, but it appears at present that several species of *Paradoxides* s.l. have a restricted distribution in Wales, Scandinavia and the eastern Appalachians, and the nature of possible barriers to their migration is not clear. Lochman-Balk and Wilson's (1958) concept of an 'extra-cratonic euxinic' realm circumscribing much of Laurentia was found unacceptable by Palmer (1973, p. A125) and the strata in question are probably best regarded as the products of poorly ventilated basins fringing the Iapetus portions of Gondwanaland and Baltica.

Middle Cambrian trilobites of Wales, Baltica, the Avalon Platform and southern New Brunswick appear at first to bear little relationship to those of the Mediterranean region, including Spain where Sdzuy (1971) established a succession of 'niveles' (or horizons), later elevated to zonal status (Sdzuy, 1972; Palmer, 1979), in ascending order as follows: *Paradoxides (Acadoparadoxides) mureroensis; Conocoryphe ovata; Acadolenus; Badule-*

sia; Pardailhania; Solenopleuropsis. There are, however, definite faunal links between the two regions and apparent differences probably reflect lateral changes of facies. One of the most widespread trilobites in the *Badulesia* Zone is *Badulesia tenera*, described first from New Brunswick but since reported from the Avalon Platform (with *Paradoxides bennettii*), Spain, Morocco, Germany and south central Turkey. All these occurrences appear to be in shales and the species has not been reported from limestones or dolomites.
Solenopleuropsis, often abundant in shales of its named zone in Spain and southern France, occurs uncommonly in thin calcarenites as far east as southeastern Turkey (Dean, 1982), then part of Gondwanaland including the Arabian Platform. In dark, anoxic mudstones of the Menevian 'Group' of south Wales *Solenopleuropsis variolaris* occurs uncommonly with *Paradoxides davidis*, an assemblage matched in the Manuels River Formation of eastern Newfoundland (Howell, 1925).
Clarella venusta, a centropleurinid from the *Paradoxides hicksii* 'Zone' of eastern Newfoundland, is recorded also from south Wales, and *C.* cf. *venusta* from southern France (Courtessole, 1975). *Centropleura*, founded on a Swedish species, has a notably wide distribution from western Europe to Siberia, as well as in Nevada (Palmer and Stewart, 1968). *Clarella*, *Centropleura* and the related *Anopolenus*, like *Paradoxides*, fall within Palmer's (1973) category of 'open ocean region' trilobites, inhabiting deeper water and therefore capable of migration over long distances along the continental margins.
By contrast the Middle Cambrian rocks of Laurentia comprise platform carbonates, deposited under warm marine conditions (Spjednaes, 1981), that contain benthic trilobites unknown from the Gondwanaland side of Iapetus. As noted earlier, no Middle or Upper Cambrian strata are known from northwest Scotland, where Lower Cambrian rocks with *Olenellus* are overlain disconformably by early Ordovician dolomites containing trilobites of the Bathyurid Province (see later). The disconformity involving the absence of Middle and Upper Cambrian at Durness extends into east Greenland and Spitsbergen; at Smøla Island, western Norway, early Ordovician gastropods of Durness – North American type occur in what Størmer (1967) termed the 'Limestone-Dolomite Belt – the western Eugeosyncline', originally part of the Laurentia cratonic margin.
In the Western Platform of Newfoundland and on the adjacent mainland, the Lower Cambrian contains both *Olenellus* and archaeocyathid reefs (Whittington and Kindle, 1969). The late Middle and early Upper Cambrian of the Port-au-Port Peninsula, near the southwestern end of the Western Platform, include lithologies and trilobites (Palmer, 1969, p. 142) that closely match those of central Montana and indicate the inner detrital belt. They must have been separated by a broad carbonate bank from the north part of the Western Platform, where contemporaneous rocks and faunas match those of the outer detrital belt, distributed around the margins of Laurentia from the western Appalachians to Nevada and Alaska.
In the central Appalachians Cambrian fossils are known from only one locality, on Dunnage Island, off the north coast of Newfoundland and within the Central Mobile Belt. Trilobites from a bed of limestone probably low in the Dunnage Formation were recorded by Kay and Eldredge (1968) as *Kootenia* and *Bailiella*, genera considered by them as characteristic of, respectively, 'Pacific' and 'Atlantic' faunas, with a 'mixing of forms from the two provinces' in central Newfoundland. The material is now being studied by Dr. W.H. Fritz, Geological Survey of Canada, Ottawa, who has kindly confirmed (personal communication) that *Kootenia* sp. and *Bailiella* sp. are present, but that the specimens are too distorted for more precise determination. The palaeogeographic implications of the assemblage should not be overestimated. *Kootenia* is widespread, having been recorded from western North America to Asia, and includes species such as *K. lakei*, from Comley, Shropshire. *Bailiella* was founded on material from the Saint John area, New

Brunswick, where the Middle Cambrian rocks and faunas display marked Welsh-Scandinavian affinities and represent a continuation of the Avalon Platform of Newfoundland. Several species of *Bailiella* have been described from Sweden (Westergård, 1949) and the genus is recorded from southern France (Courtessole, 1973) and China (Lu and others, 1965) so that its value as a provincial index must be considered dubious. In palaeogeographic terms the Dunnage assemblage could well be placed near the Gondwanaland side of Iapetus.

Ordovician

The rocks of the type Ordovician were deposited in a subsiding area once popularly known as the Welsh Lower Palaeosoic Geosyncline but nowadays termed Welsh basin or trough. Some regression at the end of the Upper Cambrian (Merioneth Series) is implied by widespread (though not invariable) stratigraphic discontinuity at this level in Wales and England, and was followed by regional transgression at the base of the Tremadoc (see later). Further regression at the end of the Tremadoc was followed in turn by transgression in the Arenig (including the so-called 'Arenig overstep' in northwest Wales), though on a smaller regional scale than was attained later, in the Llanvirn. Three broad facies belts in the Welsh Ordovician (see Bassett, 1980 for review) range from graptolitic shales in central Wales, through offshore, mixed shelly-graptolitic facies, to shelly marginal facies. Volcanic rocks are known from the Tremadoc in Wales (Kokelaar, 1979), and are extensively developed there from the Arenig to middle Caradoc.

Tremadoc Series (Fig. 5)

Recent researches directed towards an internationally agreed Cambrian/Ordovician boundary have focussed attention on the Tremadoc Series, a unit that will almost certainly be considered as lowest Ordovician. The Tremadoc is marked by widespread marine transgression and its base coincides with the almost worldwide appearance of the dendroid graptolite *Rhabdinopora flabelliformis*, a species previously known (Erdtmann, 1982, p. 27) as *Dictyonema flabelliforme*. *R. flabelliformis* is of particular importance as it provides both a cosmopolitan datum horizon and a link between the two sides of Iapetus, then sufficiently wide to constitute a barrier to benthic faunas.

The best preserved, abundant Tremadoc faunas in England and Wales are those of the Shineton Shales in Shropshire, where successive zones of *R. flabelliformis, Clonograptus tenellus,* and *Shumardia pusilla* were recognised (Stubblefield and Bulman, 1927). A still higher zone, of *Angelina sedgwickii*, is known only from Portmadoc, on the northwest coast of Wales, and is abruptly cut out by the angular unconformity beneath the overlying Arenig strata (the so-called Arenig overstep).

There is a good deal of evidence (Dean, 1976, p. 243; Owens and others, 1982, p. 3) for palaeogeographic continuity between England, Wales and certain areas now part of the eastern Appalachians, as well as related regions of South America, particularly Argentina and Bolivia. The McLeod Brook Fm. of Cape Breton Island, Nova Scotia includes lithologies remarkably like those of the Shineton Shales and contains such typical Shineton trilobites as *Asaphellus homfrayi* and *Shumardia pusilla* (Hutchinson, 1952). In eastern Newfoundland the Clarenville Formation of Random Island comprises grey shales, generally assumed to be of Tremadoc age, containing trilobites that are sometimes locally abundant but show little diversity. The fauna includes *Beltella, Angelina?, Araiopleura* and *Parabolina argentina*, the last-named species described first

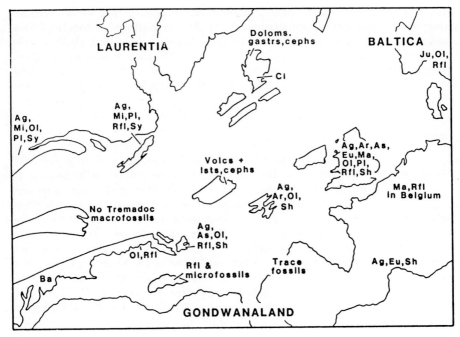

Fig. 5
Tremadoc Series: Ag, Agnostidae; Ar, *Araiopleura*; As, *Asaphellus*; Cl, *Clelandia*; Eu, *Euloma*; Ju, *Jujuyaspis*; Ma, *Macropyge*; Mi, *Missisquoia*; Ol, Olenidae excluding *Jujuyaspis* and *Plicatolina*; Pl, *Plicatolina*; Rfl, *Rhabdinopora flabelliformis*; Sh, *Shumardia pusilla* and related species; Sy, *Symphysurina*.

from Argentina, where it has been used as a zonal index in the lower Tremadoc. Of particular relevance to these faunas is Rushton's (1982) recent description of a newly discovered section at Bryn-llin-fawr, North Wales, where shales of the *Acerocare* Zone, topmost trilobite zone of the Swedish Upper Cambrian, are succeeded conformably by basal Tremadoc shales with *R. flabelliformis*. *Acerocare* itself is not present but its named zone is indicated by another Swedish olenid *Parabolina heres*. The *Acerocare* Zone at Bryn-llin-fawr contains also *Araiopleura*, *Beltella*, *Parabolinella*, *Shumardia*, *Niobella* and, at a neighbouring section, *Plicatolina kindlei*. If the first appearance of the *R. flabelliformis* species-group is used as a datum, then several trilobite genera and species once considered to be exclusively Tremadoc in age are known now to be in part of Upper Cambrian age. Rushton considered *P. argentina* to be a possible junior subjective synonym of *P. frequens*, a species described from the Leimitz-Schiefer (low Tremadoc) of Bavaria; if so, *P. frequens* had a notably wide distribution in the olenid facies around and on the Gondwanaland platforms and its vertical range spanned the Cambrian-Ordovician boundary.

On the other (= north) side of Iapetus *Plicatolina kindlei* was described first (Shaw, 1951) from Highgate Falls, Vermont, where the species occurs at a small section only a few km east of the 'Appalachian Front'. The succession there comprises thin beds of dark shale (mostly lacking macrofossils), alternating with thin siltstone bands, possibly turbiditic, in which disarticulated trilobite remains are often locally abundant and include numerous agnostids together with *Missisquoia typicalis*. Since its introduction by

Winston and Nicholls (1967) as the lowest zone of the Ordovician in central Texas, the distribution of the *Missisquoia* Zone around the periphery of Laurentia, from the Yukon, through Alberta, Oklahoma and Texas, Vermont and western Newfoundland, has been abundantly documented and the zone has acquired considerable support as a potential indicator of a basal Ordovician age.

The *Missisquoia* Zone is generally shown immediately below the *Symphysurina* Zone, long regarded as basal Ordovician by North American geologists, but the two eponymous genera occur together at Highgate Falls and elsewhere in North America, both east and west. The Cow Head area of northwest Newfoundland, described by Whittington and Kindle (1969) and later in greater detail by Fortey, Landing and Skevington (1982), provides a link between Laurentia and Gondwanaland in the latest Cambrian and earliest Ordovician. Palaeogeographically situated on the inner part of the continental slope, in a position analogous to the Cambrian outer detrital belt of Palmer (1969), the area exhibits extensive developments of the so-called 'Cow Head Breccias', conglomerates composed of clasts from the adjacent Laurentian carbonate platform. The conglomerates thin downslope and pass laterally into thin-bedded limestones containing uncommon trilobites essentially contemporaneous with those of the associated limestone clasts. On the basis of graptolites from interbedded shales Fortey and others (1982) demonstrated that *Missisquoia* and *Symphysurina* appear slightly before the first *Rhabdinopora flabelliformis*, a situation reminiscent of that in North Wales, where *R. flabelliformis* appears a little later than certain trilobites considered once to be typically Tremadoc.

Among trilobites of Tremadoc age in the platform carbonates around Laurentia is the genus *Clelandia*, a diminutive form founded on a species from New York State but since recorded from both eastern and western North America (Norford, 1969). In terms of the lettered zonal scheme established by Ross (1949) for the Ordovician of cratonic North America *Clelandia* occurs typically in zones A and B, equated with part of the lower Tremadoc (Ross and others, 1982).

In the Girvan district, just north of the Southern Upland Fault in southern Scotland, a single clast of weathered limestone from the Benan Conglomerate yielded cranidia of *Clelandia* and a pygidium determined as *Leiostegium?* (or *Symphysurina?*) by Rushton and Tripp (1979). The conglomerate forms part of the Barr Group, of lowest Caradoc age, and the clast with *Clelandia* is thought to have been derived from shallow-water limestone on the southeast side of the Dalradian trough. No comparable fauna is known elsewhere in the British Isles.

Tremadoc trilobites from Oaxaca, southern Mexico, described by Robinson and Pantoja-Alor (1968) are of particular interest as they include, inter al., *Angelina*, *Asaphellus*, *Pharostomina* and *Shumardia*, all of which are known from Welsh, English or European faunas. Whittington and Hughes (1974) placed Oaxaca on the Gondwanaland side of Iapetus and within their Olenid-Ceratopygid Province, a region comprising also Scandinavia, where the Tremadoc (including the *Ceratopyge* Limestone) is more calcareous than that of England and Wales and may reasonably be placed at a lower latitude, a position in keeping with the temperate to subtropical conditions postulated for the Swedish Ordovician by Jaanusson (1972).

An interesting component of the Lower Tremadoc fauna at Oaxaca is *Saukia*, a widespread trilobite in North America and eastern Asia, so that it should merit inclusion with the 'Pacific faunas'. The oversimplification inherent in the use of such a term is further emphasised by the occurrence of *Saukia* in southeastern Turkey and of Upper Cambrian trilobites of Asiatic affinities in Turkey, southern France and Spain, that is say in a region occupied during the Lower and Middle Cambrian by faunas of 'Mediterranean' or Gondwanaland type.

Arenig Series, Llanvirn Series and Llandeilo Series (Figs. 6, 7)

Arenig trilobites in Wales, northern England and southeasternmost Ireland (Tagoat area) show strong affinities with those of Europe, northwest Africa, the Mediterranean region and still farther east, and provide evidence for the palaeogeographic continuity of Gondwanaland. The Arenig of the Gondwanaland margins marks an important time in the history of the calymenid (s.l. including homalonotid), dalmanitid and trinucleid trilobites. These groups are not encountered in Laurentian faunas until later in the Ordovician, when migration was possible across a narrowing Iapetus, and their early evolution took place in a region of largely clastic sedimentation, particularly of muds and silts, in cool water at high latitudes (Spjeldnaes, 1981). Coincident with these Gondwanaland habitats we find, on the one hand, the apparently rapid development of large schizochroal eyes and a proparian facial suture in the dalmanitids (a successful combination destined to persist until the Upper Devonian) and, on the other, the apparent or near blindness of the trinucleids and raphiophorids, related groups with opisthoparian suture, modified in the trinucleids. The basic trinucleid plan was established early in the Arenig and continued throughout the Ordovician until the demise of the group in the late Ashgill. The nature of the trinucleid cephalic fringe remains enigmatic but the bifurcating interradial ridges in the Llanvirn genus *Stapeleyella* strongly resemble the alimentary caeca seen on some agnostid cephala (Öpik, 1961) and may have had a similar function.

Although benthic Ordovician faunas in England and Wales have been considered, for example by Hammann and others, 1982, to differ considerably from those of the Mediterranean region, nevertheless the presence in the Arenig of both regions of *Neseuretus*, *Hanchungolithus* and cyclopygids suggests that a marine connection existed. The value of *Neseuretus* as an 'index' genus in the Gondwanaland Ordovician has long been apparent (Dean, 1966, 1967; Fortey and Morris, 1982). But the genus has not been found in the rich, varied Bohemian faunas and its stratigraphic range, from Arenig to Llandeilo, followed by a related subgenus in the Caradoc, is less than that of the odontopleurid *Selenopeltis*. The latter, nominal genus of Whittington and Hughes's Selenopeltis Province, is recorded from the Arenig to Ashgill but has not been found throughout that range in any one area.

Certain Arenig facies have a remarkably wide distribution in Gondwanaland and may be useful in palaeogeographic reconstructions for both Caledonide and Mediterranean regions. The Grès armoricain, or Armorican Quartzite, described first from northwest France, is developed also in Spain, Portugal and Morocco (Spjeldnaes, 1961) and has been recorded from Bolivia (Seilacher, 1963). It was recognised in eastern Newfoundland as long ago as 1914 when Van Ingen (1914) noted inarticulate brachiopods like those of the 'Armoricains Grit' in sandstones of the Bell Island 'Group' (Rose, 1952); the succeeding Wabana 'Group' contains the distinctly Gondwanaland trilobites *Neseuretus* and *Ogyginus* (Dean and Martin, 1978). The Wabana sedimentary iron ores of Bell Island are but one of a number of comparable deposits located in the western European portion of Gondwanaland, where they range in age from Tremadoc to Caradoc. Whether made by trilobites or, as has been suggested, by another group of arthropods, the trace fossil *Cruziana furcifera* occurs in sandstones and sandy shales not only in eastern Newfoundland but also in Wales, Iberia, northwest Africa and parts of the Near and Middle East. The facies probably represents a near-shore environment around part of the Gondwanaland periphery and is not recorded from either Baltica or Laurentia.

The overall composition of Llanvirn trilobite faunas in much of Gondwanaland is similar to that in the Arenig, reflecting continuing deposition of shales, mudstones and sand-

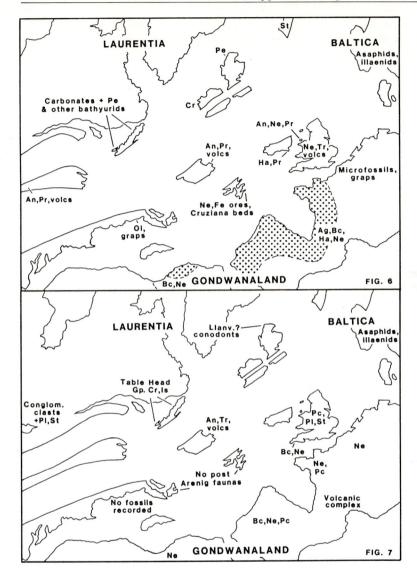

Fig. 6
Arenig Series: Ag, Agnostidae; An, *Annamitella*; Bc, *Bathycheilus*; Cr, *Carolinites*; Ha, *Hanchungolithus*; Ne, *Neseuretus*; Ol, Olenidae; Pe, *Petigurus*; Pr, *Productorthis*; Tr, Trinucleidae, excluding *Hanchungolithus*.
Dotted area denotes distribution of Armorican Quartzite.

Fig. 7
Llanvirn and Llandeilo Series (undifferentiated): An, *Annamitella*; Bc, *Bathycheilus*; Cr, *Carolinites*; Ne, *Neseuretus*; Pc, *Plaesiacomia*; Pl, *Platycalymene*; Is, Isocolidae; Pr, *Productorthis*; St, *Stapeleyella*; Tr, Trinucleidae excluding *Stapeleyella*.

stones, sometimes, as in parts of Wales, interbedded with volcanic rocks. The commonest groups are trinucleids (particularly *Stapeleyella* and *Bergamia* — see later), asaphids, cyclopygids and calymenids (including *Platycalymene* and *Flexicalymene* — see later); illaenids appear for the first time in the region.

No post-Arenig Ordovician rocks or fossils are known from eastern Newfoundland and evidence for the southward extension of Gondwanaland strata and shelly faunas is sparse. A supposed occurrence in eastern Nova Scotia is inconclusively dated, and the Arenig age of the Saint John Group in southern New Brunswick is founded mainly on graptolites. The thick clastics of the Meguma Group in southern Nova Scotia have been interpreted as part of northwest Africa (Schenck, 1971); Tremadoc and Arenig are demonstrable by means of microfossils, but the only macrofossils yet known are dendroid graptolites.

In the remainder of the eastern Appalachians the sole evidence of Gondwanaland faunas is in the basement rocks of Florida, where *Colpocoryphe exsul* Whittington (1953), from a shale/sandstone succession of Llanvirn or Llandeilo age, is now assigned (Dean, 1976) to *Plaesiacomia* and belongs to a genus especially common in Iberia and France.

From the Arenig onwards and for much of the Ordovician, the western part of Baltica was the site of mud and silt deposition with subsidiary carbonates and sands, so that both shelly and graptolitic faunas occur. According to Jaanusson (1972) the climate was temperate, occasionally subtropical, conclusions endorsed by the palaeogeographic map of Spjeldnaes (1981). The most abundant trilobite groups include asaphids (the region is part of Whittington and Hughes's Asaphid Province), illaenids and raphiophorids *(Ampyx, Lonchodomas)* with subsidiary cheirurids, and show little evidence of close links with the British or European parts of Gondwanaland, though the graptolitic shales of the lower Arenig in southern Norway contain what Størmer (1967, p. 196) described as a British-Scandinavian fauna. Trinucleids are unknown in Baltica before the Llanvirn.

The near equatorial position of much of Laurentia in the early Ordovician, with extensive deposition of platform carbonates around the margins, led to the establishment there of a biostratigraphic scale very different from that in Europe and Baltica (Williams and others, 1972), with consequent problems of detailed correlation that are not yet resolved. The most recent review (Ross and others, 1982) employs the following Series, in ascending order: Ibex (replaces Canadian of earlier usage); Whiterock; Mohawk; and Cincinnati. The two last-named are subdivided into Stages, but no generally accepted stage names are available for the first two. In broad terms: Ibex Series equals Tremadoc plus lower Arenig; Whiterock Series equals upper Arenig plus Llanvirn, Llandeilo and basal Caradoc; Mohawk Series equals most of the Caradoc; and Cincinnati Series equals topmost Caradoc plus all the Ashgill.

From Arenig to Llandeilo, Laurentia was the site of Whittington and Hughes's Bathyurid Province, and trilobites of that family persisted into the Caradoc on the North American craton; elsewhere, as noted later, their distribution is more complex and may involve different regions and facies. As during the earlier Phanerozoic, the seaward side of the carbonate platform was marked by a transition through interbedded carbonates and muds into, finally, dark muds, a lateral sequence of sediments and associated faunas reviewed by Fortey (1975) and by Shaw and Fortey (1977).

The widespread limestones and dolomites of North America include the Beekmantown Group of Pennsylvania, eponymous unit of a once much-used but now redundant chronostratigraphic term. In western Newfoundland similar lithologies constitute the St. George Group, in which occur the bathyurid *Petigurus nero* and gastropods found also in east Greenland. The same assemblage is recorded from the Durness Limestone Group of north-

west Scotland, and the limestones of Smøla Island, near Trondheim, western Norway, belong to the same set of deposits (Størmer, 1967).
The discovery of the Laurentian trilobite *Clelandia*, of lower Tremadoc age, near Girvan, in a conglomerate clast derived from a limestone source north of that area was noted earlier. Further evidence of Laurentian trilobites, in this instance of lower Arenig age, in central Scotland comes from a bed of silicified limestone in the Highland Boundary Complex recorded by Curry and others (1982), according to whom the rocks form part of a carbonate sequence that may extend at depth much farther into the Midland Valley of Scotland. Genera present include, inter al.: *Bolbocephalus*, a bathyurinid found in western Newfoundland; *Punka*, a bathyurellinid from western Newfoundland and Spitsbergen; *Strotactinus*, a pliomerid known from south Ontario and from limestones and clasts in southern Quebec; and *Sycophantia*, a cheirurid recorded only from Spitsbergen. The assemblage would be at home at or near the outer edge of the Laurentia carbonate platform, in a position approximating to Palmer's (1973) outer detrital belt of the Cambrian.
Fluctuations in the position of the Ordovician coastline of Laurentia are demonstrable in western Newfoundland where the St. George Group is succeeded, with basal unconformity and often associated lead-zinc mineralisation, by the Table Head Formation, whose alternating dark shales and thin, dark limestones often show evidence of contemporaneous slumping down the inner part of the continental slope, and pass upwards into unnamed sandstones and shales. The rich, varied trilobites of the autochthonous Table Head Formation at Table Point described by Whittington (1965) were shown by him to be about lower Llanvirn in age and were said to be of 'Whiterock' type. Contemporaneous platform carbonates, though not exposed in the same area, are represented by one giant boulder in a conglomerate of the allochthonous succession at Lower Head (Whittington, 1963) and by clasts in the Mystic Conglomerate of southern Quebec, both of which contain brachiopods and trilobites 'strikingly similar' to those from contemporaneous rocks in Nevada (Ross and Ingham, 1970, p. 403). The Table Head trilobites include, inter al., agnostids, raphiophorids, olenids, pliomerids, cheirurids, remopleuridids, asaphids and nileids, and constitute an assemblage that has sometimes been regarded as a standard for interpreting the term Whiterock. The original concept of a Whiterock Stage was based by Cooper (1956) on a succession of brachiopod faunas in Nevada, and the term, elevated to the status of a Series occupying about one quarter of the Laurentian Ordovician, has recently been discussed by Ross and others (1982). It is evident from Ross and Ingham's (1970) discovery in the Albany Mudstones, of Caradoc age, near Girvan, Scotland, of a trilobite assemblage that would customarily have been termed 'Whiterock' that the latter name should be used with discretion. Ross and Ingham preferred to introduce 'Toquima-Table Head faunal realm' to denote a facies peripheral to the margins of Laurentia; some of the contained genera may have a long vertical range (including Arenig to Llandeilo) and correlation of faunas should, ideally, be based on species. Further evidence of this comes from the Mayo-Galway area of western Ireland (reviewed by Williams and others, 1972, p. 54) where a Laurentian trilobite assemblage was thought to be of Llanvirn age on the basis of its 'Whiterock' (presumably Table Head) aspect; graptolites from adjacent shales show the age to be upper Arenig. An interesting member of the western Irish fauna is the telephinid trilobite *Carolinites*, a probably pelagic genus with very large eyes that was founded on an Arenig species from Tasmania but is not uncommon around the periphery of Laurentia. *Carolinites* was capable of migration in regions of deeper water and some of its species, like certain agnostids and olenids in the Cambrian, are almost cosmopolitan (Fortey, 1975).

Trinucleid trilobites, so well represented in the early Ordovician shales and mudstones of Gondwanaland, are notably absent from contemporaneous shales around Laurentia. A striking exception is a single specimen of *Stapeleyella* sp. from a limestone clast in a probably Llandeilo (*N. gracilis* Zone) conglomerate at Quebec City, eastern Canada (Hughes and others, 1975, p. 559; Dean, 1976, p. 244). All known datable occurrences of *Stapeleyella* are in Llanvirn rocks, including those of Wales, Shropshire and Bohemia, and a similar age is likely for the Canadian specimen. The type species of *Trinucleus*, *T. fimbriatus*, was described from black, graptolitic shales of Llandeilo age in central Wales and, as for the olenids, evidence for its mode of life is equivocal. The euxinic environment suggests that the species may have been pelagic; on the other hand it may been adapted to a partially benthic life in anoxic conditions. *Stapeleyella* is placed in the same subfamily, Trinucleinae, by Hughes and others (1975) and its mode of life may have been comparable; both genera, like the widespread trinucleinid *Tretaspis*, have a distinctly segmented glabella and subspherical frontal glabellar lobe, and lack an occipital spine.

Many limestone clasts containing trilobites of Llanvirn age and 'Whiterock' (i.e. Toquima-Table Head) type have been found in the Mystic Formation of south Quebec, a short distance east of the 'Appalachian Front' (Dean, 1976, p. 232). One clast yielded a single cranidium of *Platycalymene*, another genus with type species from black shales in central Wales and therefore adapted to an anoxic environment, if not actually pelagic. Ross and Ingham (1970, p. 396) listed *Platycalymene* from the Albany Mudstones (Caradoc) at Girvan, Scotland, in a fauna that contains numerous Toquima-Table Head genera.

Central Appalachian and Related Pre-Caradoc Faunas (Figs. 5—7)

Ordovician shelly faunas in the Central Volcanic Mobile Belt of Newfoundland and analogous parts of New Brunswick and Maine pose a number of problems in terms of Iapetus palaeogeography. Neuman (1972) regarded lower Ordovician brachiopods there as having inhabited the fringes of a series of volcanic island arcs. An earlier reconstruction (Dean, 1976, fig. 6) showed the Central Belt and its faunas in an intermediate position in Iapetus, but they could be more appropriately placed closer to Gondwanaland, at least in the Arenig and perhaps in the Llanvirn and Llandeilo. Tremadoc evidence is sparse and equivocal, and by the Caradoc, when volcanic activity had ceased, the region was the site of extensive black, sometimes graptolitic mud deposition.

In the Central Belt of Newfoundland some of the Ordovician brachiopods and trilobites have much in common with Appalachian faunas from the Shin Brook Fm. of Maine, northeast U.S.A. The oldest Newfoundland assemblage, of about Arenig age, includes Baltoscandian elements such as *Metopolichas*, *Illaenus* (s.l.) and *Pseudosphaerexochus*, together with *Annamitella?*, placed in the Bathyuridae and described also from Maine (Whittington in Neuman, 1964) in company with *Nileus*, found in both the Whiterock Series of Nevada and the lower Ordovician of Baltica. Brachiopods of Whiterock type similar to those of Newfoundland occur in central New Brunswick (Neuman, 1968). The youngest trilobites from the volcanic terrain of central Newfoundland include cheirurids, encrinurids, pliomerids, trinucleids, illaenids, nileids and styginids. The assemblage exhibits closest affinities with those of the Porterfield Stage (approximately middle Llandeilo to basal Caradoc; Williams and others, 1972, fig. 2) in both the southern Appalachians and the Girvan district of Scotland; but there is in addition a strong resemblance, shown also by associated conodonts (Uyeno in Dean, 1971, p. 37), to faunas in Baltica.

In Scandinavia *Trinucleus forosi* was described first from slates low in the Hovin Group of the Hølonda area, south of Trondheim, western Norway, in what Størmer (1967,

p. 192) later termed the 'Shale-Sandstone-Lava Belt — the eastern Eugeosyncline' of the Caledonides and now included in the exotic Trondheim Nappe Complex. The species was assigned to *Stapeleyella* by Størmer (loc. cit.) and, questionably, by Hughes and others (1975, p. 559). More abundant trilobites (excluding trinucleids) and brachiopods from siltstones in the Hølonda area were revised by Neuman and Bruton (1974) and shown to be of 'Whiterock' type, of late Arenig-early Llanvirn age and North American affinities. Further description by Bruton and Bockelie (1980) envisaged the Hølonda area as having occupied part of a volcanic island arc in a back-arc basin off the margin of Laurentia, and earlier remarks concerning *Stapeleyella* in the 'Whiterock' of the Quebec area, Canada, could well apply to its occurrence in Norway.

The Ordovician faunas of Anglesey, a large island off the northwest coast of Wales, are highly relevant to the present discussion. The stratigraphy and structure of Anglesey were described by Bates (1972, 1974), who had earlier (Bates, 1968) documented trilobites and brachiopods from what he termed the 'Principal Area' in the northwest half of the island. The rock succession, which overlies the Mona Complex (Precambrian) with marked unconformity, is as follows: Carmel Fm., Treiorwerth Fm., and Nantannog Fm. (including the so-called Bod Deiniol Fm., a lenticular unit within the Nantannog Fm.). The Carmel Fm., lower Arenig, contains the trilobites *Neseuretus monensis, Merlinia selwynii* and *Monella perplexa*, the last-named nowadays considered congeneric with the *Annamitella?* species from the Appalachians noted earlier.

Monella? was recorded from the Treiorwerth Fm., the brachiopods from which were revised by Neuman and Bates (1978). Of the eighteen brachiopod taxa, six could not be identified generically, six genera were endemic to Anglesey and the Tagoat area, southeastern Ireland (see below), four genera (including *Productorthis*) were found in association elsewhere only in the central belt of Newfoundland, one *(Hesperonomiella)* was described from the Whiterock Series of western U.S.A., and one *(Skenidioides)* had such a long stratigraphic range as to be of dubious significance. Of the seven brachiopod genera in the Bod Deiniol 'Fm.', all were known from the central belt of Newfoundland and five from the upper Arenig and Llanvirn of the Baltic area.

The small outcrop of Arenig rocks near Tagoat, Co. Wexford, near the southeastern tip of Ireland, described by Brenchley, Harper and Skevington (1967) contains graptolites indicative of the *Didymograptus extensus* Zone and the trilobites *Bergamia* sp., ?*Ogyginus* sp., *Merlinia* cf. *selwynii* and the trinucleid *Hanchungolithus* aff. *primitivus*. The first three of these genera are widespread in Wales; *H. primitivus* is abundant and of stratigraphic value in the Montagne Noire, southwestern France (Dean, 1966). Brachiopods from Tagoat were identified by Bates who noted their close resemblance to faunas of similar age from Anglesey (see below).

Neuman and Bates (1978, p. 577) considered the Anglesey faunas revised by them to reinforce the concept of a Celtic Province, introduced earlier by Williams (1973, p. 249) to accommodate the Arenig brachiopods of both Anglesey and the Tagoat area, and they extended that province to accommodate the central belt of the Appalachians in Newfoundland, New Brunswick and Maine. My view is that, on the contrary, the term Celtic Province should be abandoned. The Anglesey and Tagoat faunas demonstrate a lateral overlap of undoubted Gondwanaland trilobites such as *Neseuretus* and *Hanchungolithus* with genera that inhabited not only fringes of volcanic islands such as existed during the early Ordovician in the northern Appalachians, including Newfoundland, but also, when local conditions permitted, the margins of at least part of Gondwanaland. The further geographic extension of 'Central Appalachian' faunas is not clear, but *Annamitella?* may be congeneric with both *Bathyuriscops* from Kazakhstan and *Proetiella* from Argentina (for discussion, see Whittington in Neuman, 1964, p. E30; Dean, 1976, p. 236). The Precordillera of western Argentina (Harrington and Leanza, 1957) contains Ordovician rocks from upper Arenig to upper Caradoc in age totalling more than 4000 metres. In dark limestones of Llanvirn age are found *Annamitella?* (as *Proetiella*), illaenids, pliomerids and remopleuridids, and associated brachiopods include *Productorthis*. The type species of *Annamitella, A. asiaticus* from Indochina, requires redescription, but if all these records of *Annamitella?* are congeneric, then we are dealing with an unusually widespread early Ordovician representative of the Bathyuridae outside Whittington and Hughes's (1972) Bathyurid Province.

Llandeilo to basal Caradoc volcanic rocks and associated sediments in central Newfoundland are overlain by black shales that lack shelly fossils but contain early Caradoc graptolites exhibiting affinities with the Southern Uplands of Scotland (Toghill in Dean, 1971). Black shales occur also in central New Brunswick but vulcanicity there extended into the mid-Caradoc, and associated limestone lenses contain a few trilobites of Porterfield aspect and conodonts of Baltoscandian affinities (Dean and Uyeno in Skinner, 1974). Post-volcanic black shales of Caradoc age are widespread not only in the central Appalachians but also in Wales, and even extend across eastern parts of the present Canadian Shield. Spjeldnaes (1981, p. 218) postulated that anoxic deposits of this type were formed when 'polar ice caps were absent and the climate was warm with diffuse zoning'.

Early Ordovician generic assemblages of brachiopods and benthic trilobites described as Whiterock or Toquima-Table Head faunal realm fit conveniently into the circum-Laurentia pattern shown by Ross and Ingham (1970); they occupy a position outside the carbonate platform corresponding to the upper (or inner) slope of Shaw and Fortey (1977). That occasional examples of possibly pelagic trilobites of Gondwanaland type should also be found there is, perhaps, not so surprising, but it is less easy to explain the mixing of typical Gondwanaland benthos in the Arenig of Wales and southeast Ireland with *Annamitella?* and brachiopods that occur on the inner slope of Nevada. Some genera clearly have a long stratigraphic range as well as wide geographic distribution, so that precise age determinations rest on identification at specific level, not an easy task in areas where tectonic distortion is common. Neuman (1972) suggested that volcanic islands provided suitable migration routes for Ordovician brachiopods in the Appalachians, and Shaw and Fortey (1977, p. 433) noted 'island-hopping' as a possible means of migration for trinucleids there. But such hypotheses are not totally satisfactory for an Iapetus whose breadth in the early Ordovician (Arenig) has been postulated as between 3000 and 4000 km (McKerrow and Cocks, 1976) and about 4000 km (Cocks and Fortey, 1982). The problem was alluded to by Bruton and Bockelie (1980) in assessing the Hølonda faunas of western Norway. It is possible that Iapetus was narrower than has been postulated, though no wholly adequate explanation is yet available, and some benthic animals, or perhaps their larval stages, may have been capable of more extensive migration than is generally appreciated.

Caradoc Series (Fig. 8)

Some of the most complex Ordovician faunal distributions in the Iapetus region are found in the Caradoc and may be related to plate movements and progressive ocean closure. The oldest Caradoc trilobites found in a shale environment in Wales and the Welsh Borders (Whittard, 1966) comprise calymenids, trinucleids and asaphids that represent essentially a continuation of indigenous Llandeilo groups. In shallow marine habitats, however, the relationships are more complex and the numerically dominant

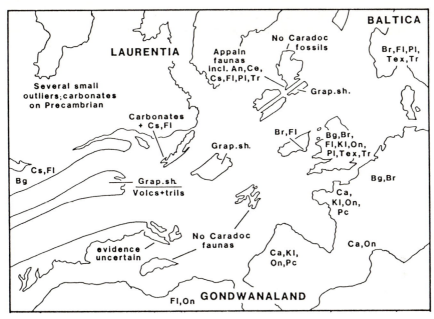

Fig. 8

Caradoc Series: Bg, *Brongniartella;* Br, *Broeggerolithus;* Ca, *Calymenella;* Ce, *Ceraurinella;* Cs, *Ceraurus;* Fl, *Flexicalymene;* Kl, *Kloucekia;* On, *Onnia;* Pc, *Plaesiacomia;* Pl, *Platycalymene;* Tex, *Toxochasmops extensa;* Tre *Tretaspis.*

brachiopods (Williams, 1973) include Baltic, Scandinavian and Bohemian elements in addition to the more common indigenous groups; similar criteria apply to the trilobites. The late Llandeilo and earliest Caradoc coincide with the so-called *Nemagraptus gracilis* Transgression, involving almost world-wide distribution of the nominal graptolite. The transgression must have been instrumental in changing established faunal patterns, and its effects extended into the early part of the succeeding *Diplograptus multidens* Zone.

The Longvillian Stage of Bancroft (1929, 1933) was divided by him into Lower and Upper substages in its type area of south Shropshire. More recently Hurst (1979) restricted Longvillian to Bancroft's Lower Longvillian and gave a new name, Woolstonian, to the Upper Longvillian. This usage is followed here, bearing in mind that it can be difficult to distinguish between Lower and Upper Longvillian in facies other than those of the type area. Similar comment applies to the succeeding Marshbrookian, which is little known, or little represented, elsewhere. Williams (1976) showed that of the numerous brachiopods in the type Longvillian and Woolstonian, on the edge of the English Midland Platform, few are found in the deeper water, eugeosynclinal environments west of the Linley-Pontesford and Church Stretton fault systems, structural lineaments that mark the boundary between platform and trough in the Welsh Lower Palaeozoic and can be traced into South Wales.

On the other hand the trinucleid trilobite *Broeggerolithus nicholsoni* and the homalonotid *Brongniartella,* including especially *B. minor,* constitute an assemblage that is widely distributed in North Wales (Gelli-grîn Formation, Llanbedrog Mudstones) and northern England (Drygill Shales, Corona Beds), particularly in a mudstone facies. It is found, also in mudstones, in the Slieveroe area of southeast Ireland (Brenchley and others, 1977) as well as in southern Belgium (writer's unpublished results). In terms of present palaeogeographic reconstructions these trilobites could have migrated along the northern margin of the Midland Platform, or even across the platform if one takes into account the borehole record from Kent, southeast England, of a homalonotid and ?trinucleid in association with Caradoc microfossils (Lister and others, 1969). The dalmanited *Kloucekia,* found in the Longvillian and Woolstonian of Shropshire after being absent since the Harnagian, occurs in shallow water deposits at several places in north Wales, though not recorded from the Woolstonian there. The genus suggests links with faunas in Bohemia and southwestern Europe, though possible migration routes are not clear and *Kloucekia* has not yet been found in Belgium.

A significant trilobite in the Longvillian of North Wales is the pterygometopid *Chasmops cambrensis,* a species which, though not found in Shropshire, indicates links with Baltica, where *C. conicophthalma* is characteristic of the Lower Chasmops Shale and Limestone (Etages 4bα and 4bβ) in the Oslo region (Størmer, 1940). Further evidence of connection between Baltica and Wales derives from another pterygometopid *Estoniops,* found in the Woolstonian of north Wales and northern England (Cross Fell inlier). Mudstone faunas of Longvillian/Woolstonian type at Ringsaker, Norway (Dean, 1960) do not include trinucleids, but according to Hughes and others (1975, pp. 580—1, 588) *Broeggerolithus* of Anglo-Welsh type reached Sweden (Dalarne) by the middle Caradoc and persisted in Norway until the end of the Caradoc.

The Woolstonian of Shropshire contains the first British representatives of *Toxochasmops* [*Chasmops*] *extensa,* a species that becomes progressively more abundant there in the Marshbrookian and Actonian Stages and is characteristic of the Upper Chasmops Shale and Limestone (stages 4bγ and 4bδ) in the Oslo area, Norway (Størmer, 1940, 1967). Trilobites, including the raphiophorids *Ampyxella* and *Lonchodomas,* the olenid *Triarthrus,* and the pterygometopid *Calyptaulax,* from Actonian and Onnian mudstones in Shropshire show a striking resemblance to contemporaneous Norwegian and Swedish faunas. On the other hand the trinucleid *Onnia,* characteristic of the Onnian, highest Caradoc stage, is thought to have evolved in southern Europe and may represent an immigrant to the Welsh area.

Such distribution patterns of benthic faunas can be interpreted as evidence of the increasing proximity of the Anglo-Welsh portion of Gondwanaland and Baltica, beginning in the Longvillian or the Woolstonian. But the Caradoc was also a time when marine connection between England, Wales and Laurentia can first be demonstrated on the basis of benthic trilobites. *Brongniartella,* represented by *B. trentonensis,* occurs in both Pennsylvania (Whitcomb, 1930) and Virginia, where it is accompanied by the harknessellid brachiopod *Reuschella,* described first from, and abundant in, the Shropshire Caradoc. The age of these American occurrences is about Longvillian or Woolstonian.

In the Port-au-Port Peninsula, part of the Western Platform of Newfoundland, platform carbonates in the lower part of the Long Point Group contain, in addition to trilobites of Laurentian type, rare *Flexicalymene* of Anglo-Welsh affinities (Dean, 1979). The age in North American terms is low Mohawk Series, about Blackriveran Stage, and corresponds to early Caradoc, that is to say about the same level as the widespread marine transgression in the Welsh area. The occurrence is also of interest in that the limestones are transgressively unconformable upon, and so give an upper age limit to the emplacement of, Taconic nappes in western Newfoundland (Rodgers, 1965). Other noteworthy records of calymenids in eastern North America are from the 'Trenton Limestone' (? about Longvillian) of New York State (Ross, 1967) where *Flexicalymene senaria* and *Gravicalymene* show, again, affinities with groups whose previous history lay in Gondwanaland.

Evidence of faunal interchange between Laurentia and Gondwanaland is not confined to the trilobites, and Spjeldnaes (1981, p. 223) has noted the invasion of Europe by American brachiopods in the mid-Caradoc.

Ashgill Series (Fig. 9)

In certain respects the Ashgill was a time of contrasts: on the one hand widespread, richly fossiliferous silts and muds, with occasional carbonate mud-mounds in which trilobites were locally common; on the other, a well-developed glaciation at the end of the epoch, centred on an extensive icecap at the Ordovician south pole that was the cause of large scale faunal extinctions. The regional picture, as far as the Caledonides are concerned, is dominated by the narrowing of Iapetus to permit an interchange of benthic faunas already heralded in the Caradoc of the Welsh area.

In Whittington and Hughes's (1972) interpretation of Ashgill trilobite distribution the number of faunal provinces was reduced to two, Remopleuridid and *Selenopeltis*, reflecting the closure of Iapetus to such an extent that many trilobites were able to migrate between the two opposing sides, some genera becoming almost cosmopolitan. In its type area of Cautley, northern England, the unusually complete Ashgill follows the Caradoc conformably; elsewhere in England, Wales and much of Europe the boundary is often unconformable and the sequence incomplete owing to local disconformities. Whittington and Hughes (1972, fig. 12) showed the Remopleuridid Province covering Laurentia, with its extensive platform carbonates (for which the more appropriate term Cincinnati Series is used), the British Isles, Baltica and part of China, but excluding southern Europe, North Africa and the Middle East. In the light of more recent discoveries this interpretation requires some modification, quite apart from the profound changes in marine benthic assemblages (trilobites and brachiopods are the best documented) that coincided with the glaciation centred on the Anti-Atlas of Morocco in the latest Ashgill.

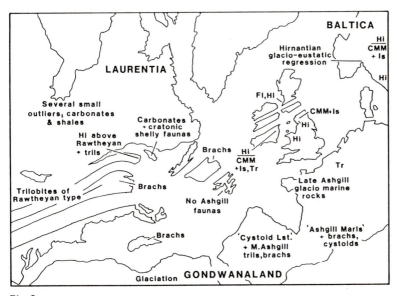

Fig. 9

Ashgill Series: Fl, *Flexicalymene;* Hi, *Hirnantia* fauna; Is, Isocolidae; Tre, *Tretaspis* CMM, carbonate mud-mounds.

One of the more common and stratigraphically useful trilobites in the Ashgill of Britain, northern Europe and Scandinavia is the trinucleid *Tretaspis*, following its appearance, in Baltica and the Welsh area, in the upper Caradoc. Its oldest known occurrence is in Virginia where *Tretaspis* was described by Whittington (1959) from both the Lantz Mills facies (buff weathering limestones) and the Liberty Hall facies (black limestone and shale) of the lower Edinburg Formation, lithologies attributed to, respectively, the upper slope and lower slope by Shaw and Fortey (1977). The rocks were correlated by Whittington with the Porterfield Stage (= middle and upper Llandeilo plus basal Caradoc in Williams and others, 1972); according to Hughes and others (1975) *Tretaspis* is not older than early Caradoc. *Tretaspis* is one of several Appalachian genera (others include *Calyptaulax*, *Ceraurinella* and *Sphaerexochus*) that greatly extended their distribution in the late Caradoc and Ashgill due to the narrowing of Iapetus. In the case of *Tretaspis*, at least, migration must have been onto and around Gondwanaland and Baltica (then in close proximity to one another) and did not extend onto the carbonate platform of Laurentia.

Migration similar to that above, but involving a very different group of trilobites, is demonstrated by the Isocolidae, typically blind or almost so. *Isocolus* was founded on *I. sjoegreni* from the Boda Limestone of Dalarne, Sweden, and the family shows a distinct preference for carbonate mud-mounds or 'reef limestones'. In the Caledonide region some degree of continuity between Baltica and the British Isles portion of Gondwanaland is suggested by the carbonate mud-mounds of Boda, Keisley (northwest England) and the Chair of Kildare (eastern Ireland), all of which have several genera and species in common and are of late, though not latest, Ashgill age. Isocolidae are well represented at Kildare by *Tiresias, Cyphoniscus* and (rarely) *Isocolus*. Of these *Tiresias* is represented in the Appalachians by *T. typus* (described first as *Holdenia*) from the Effna Limestone, noted as a shelf edge deposit by Shaw and Fortey (1977) and of Llandeilo to basal Caradoc age (Ross and others, 1982). The oldest known species of *Isocolus, I. dysdercus,* from a boulder of white, reef-like limestone of Whiterock age in the Cow Head Group of western Newfoundland, has numerous eye facets (Whittington, 1963). No eye facets are found in *I. sjoegreni* and it is possible that, as suggested by Erben (1958), the isocolids adapted 'to life in dark submarine cavities of certain reefs'.

At the family level trilobites tend to show little preference for a particular lithology, though genera and species are often subject to facies control. An interesting example of the isocolids' preference for biohermal limestones is Whittington and Orchard's (1977) record of the Kildare and Keisley genus *Cyphoniscus* from northwest Iran. Distribution of *Cyphoniscus,* found also, though rarely, in the Ashgill of southern Turkey, spans both the remopleuridid and *Selenopeltis* provinces of Whittingston and Hughes (1972); it also emphasises both the continuity of Gondwanaland and its possible proximity to Baltica in the Ashgill, with no necessity to invoke either a Proto-Tethys or a Mid-European ocean at that time.

In eastern Canada the 'European' aspect of certain Ashgill trilobites from the White Head Formation of the southern Percé region, Quebec, eastern Canada, has been appreciated for more than half a century (Schuchert and Cooper, 1930). Most of the trilobites belong both to southern Appalachian genera that had a longer history in the Caradoc, and to forms that are widespread in Britain, Scandinavia and Europe, including the carbonate mud-mounds (Boda, Keisley, Chair of Kildare) noted earlier; in Anglo-Welsh terms their age would be about Rawtheyan The Percé strata do not form part of the Laurentian carbonate platform and trilobites such as cheirurids, pterygometopids and encrinurids indicate a habitat on the upper slope, though some fluctuation of the platform margin is suggested by deeper water (lower slope) forms such as raphiophorids, trinucleids *(Tretas-*

pis) and cyclopygids. Higher in the White Head Formation occurs the late Ashgill *Hirnantia* fauna, noted below. Evidence of Ashgill faunas in central Newfoundland and corresponding parts of Main is generally sparse and derives mainly from brachiopods (literature reviewed in Dean, 1976, p. 238), but trilobites from the Pyle Mountain argillite of northeast Maine recorded by Whittington (in Boucot and others, 1964) represent the White Head Ashgill fauna of Percé and exhibit strong European affinities.

Bancroft's Hirnantian, topmost stage of the Anglo-Welsh Ordovician and founded on strata near Bala, north Wales, has received increased attention in recent years, largely as the result of work on the well-documented late Ordovician glaciation centred on the Sahara region of Africa, and its effects elsewhere including South America (review of literature in Spjeldnaes, 1981). If the base of the Silurian is drawn below the *Parakidograptus acuminatus* Zone rather than below the *Glyptograptus persculptus* Zone, the main glaciation is essentially of Hirnantian age. Agreement on interpretation of the glacial phenomena becomes more tenuous with distance from the Sahara, but the glaciomarine nature of rocks in Iberia and the presence of ice-drop tillites in Normandy appear to be generally accepted. The Hirnantian marks a notable drop in abundance and variety of benthic shelly faunas compared with the underlying Rawtheyan Stage, but the brachiopods *Hirnantia sagittifera* and *Eostropheodonta hirnantensis*, plus the dalmanitid trilobite *Mucronaspis mucronata* are widespread and form part of a '*Hirnantia* fauna', elements of which extend from the British Isles through Scandinavia, Morocco, Bohemia and Poland to China, though the local composition of the fauna may vary. According to Spjeldnaes (1981, p. 234) warm-water carbonate rocks may occur in areas of Asghill glaciation, but the claim needs clarification. In Britain many so-called Ashgill limestones are little more than lenticular, calcareous mudstones, though the 'Hirnant Limestone' of the Bala area is pisolitic. Limestones at the Chair of Kildare cited by Spjeldnaes are, in fact, of pre-Hirnantian (Rawtheyan) age, overlain by mudstones containing the *Hirnantia* fauna (Wright, 1968); but the Keisley Limestone, which has many faunal elements in common with that of Kildare, is overlain by thin limestones of Hirnantian age (Ingham and Wright in Williams and others, 1972, p. 47).

There is no detailed documentation of effects of the glaciation on Laurentian platform sedimentation, but they have been cited by Sheehan (1975) as the cause of extinctions of certain specialised, endemic stocks of brachiopods in the epicontinental faunas; the latter, according to Sheehan, were subsequently replaced by European groups that were less specialised and so better able to survive detailed changes imposed by the lowering of sealevel.

In Baltica eustatic lowering of sea-level due to the Gondwanaland glaciation was discussed by Brenchley and Newall (1980) who postulated one main regression in the Oslo region; according to them the Hirnantian glacial maximum there lasted less than one million years and involved a fall in sea-level of fifty to one hundred metres.

Whatever the precise figures, it is clear that the combination of lower sea-water temperature, possible changes in salinity, and the widespread loss of habitats resulting from regression had a profound effect on Ordovician trilobites, though whether their extinction was simultaneous in all areas has not yet been demonstrated. Families that became extinct at the end of the Ordovician include: agnostids, olenids, shumardiids, trinucleids, dionidids, pliomerids, asaphids, nileids, cyclopygids, telephinids and remopleuridids. Some of these were of benthic type and so more vulnerable to the effects of regression; others, such as agnostids and olenids, are typically associated with deeper water environments but had a long previous history, in some cases from the Cambrian, and their numbers had diminished considerably by the late Ordovician. Curiously, the isocolids are absent from the list; although probably specialised, and adapted to life in biohermal environments, their youngest representatives include *Cyphoniscus* from early Silurian bedded limestones in the Gaspé region of eastern Canada.

When the basal Silurian transgression followed the end of the glaciation, the situation had changed considerably as far as the trilobites were concerned; many Silurian genera were cosmopolitan, but the group had declined enormously in numbers and variety, and from then on played a relatively minor role in Palaeozoic faunas.

Acknowledgements

I am indebted to Dr Michael Bassett for reading the manuscript and making constructive suggestions for its improvement; to Dr Rodney Gayer for useful discussion of problems relating to the closure of Iapetus; and to Dr. W.H. Fritz, Geological Survey of Canada, Ottawa, for information on the Cambrian trilobites from Dunnage Island, Newfoundland.

References

Ahlberg, P. (1981): Ptychopariid trilobites in the Lower Cambrian of Scandinavia. In Taylor, M.E. (ed.). Short papers for the Second International Symposium on the Cambrian System 1981. U.S. Geological Survey Open-File Report 81–743, 5–7.

Bancroft, B.B. (1929): Some genera and species of Strophomenacea from the Upper Ordovician of Shropshire. Memoirs of the Manchester Literary and Philosophical Society, 73, 33–65.

Bancroft, B.B. (1933): Correlation tables of the Stages Costonian – Onnian in England and Wales. 4 p, 3 tables. Blakeney, Glos.

Bassett, M.G. (1980): The Caledonides of Wales, the Welsh Borderland, and Central England. Pp. 34–48 in Owen, T.R. (ed.). United Kingdom: Introduction to general geology and guides to excursions 002, 055, 093 and 151. International Geological Congress, Paris. Institute of Geological Sciences, London.

Bassett, M.G., Owens, R.M., and Rushton, A.W.A. (1976): Lower Cambrian fossils from the Hell's Mouth Grits, St Tudwal's Peninsula, North Wales. Journal of the Geological Society of London, 132, 623–644.

Bates, D.E.B. (1968): The Lower Palaeozoic brachiopod and trilobite faunas of Anglesey. British Museum (natural History), Bulletin, Geology, 16, 125–199.

Bates, D.E.B. (1972): The stratigraphy of the Ordovician rocks of Anglesey. Geological Journal, 8, 29–58.

Bates, D.E.B. (1974): The structure of the Lower Palaeozoic rocks of Anglesey, with special reference to faulting. Geological Journal, 9, 39–60.

Bergström, J. (1973): Organization, life, and systematics of trilobites. Fossils and Strata, 2, 1–69.

Bergström, J. (1981): Lower Cambrian shelly faunas and biostratigraphy in Scandinavia. In Taylor, M.E. (ed.). Short papers for the Second International Symposium on the Cambrian System, 1981. U.S. Geological Survey Open-File Report 81–743, 22–25.

Bergström, J., and Levi-Setti, R. (1978): Phenotypic variation in the Middle Cambrian trilobite *Paradoxides davidis* Salter at Manuels, SE Newfoundland. Geologica et Palaeontologica, 12, 1–19.

Boucot, A.J., Field, M.T., Fletcher, R., Forbes, W.H., Naylor, R.S., and Pavlides, L. (1964): Reconnaissance bedrock geology of the Presque Isle Quadrangle, Maine. Maine Geological Survey, Quadrangle Mapping Series 2, 123 pp.

Brenchley, P.J., Harper, J.C., Mitchell, W.I., and Romano, M. (1977): A re-appraisal of some Ordovician successions in eastern Ireland. Proceedings of the Royal Irish Academy, 77 B, 66–85.

Brenchley, P.J., Harper, J.C., and Skevington, D. (1967): Lower Ordovician shelly and graptolitic faunas from south-eastern Ireland. Proceedings of the Royal Irish Academy, 65 B, 385–390.

Brenchley, P.J., and Newall, G. (1980): A facies analysis of Upper Ordovician regressive sequences in the Oslo region, Norway – a record of glacioeustatic changes. Palaeogeography, Palaeoclimatology, Palaeoecology, 31, 1–38.

Bruton, D.L., and Bockelie, J.F. (1980): Geology and palaeontology of the Hølonda area, western Norway – a fragment of North America? In Wones, D.T. (ed.), The Caledonides in the U.S.A. IGCP Project 27: Caledonide Orogen 1979 Meeting, Department of Geological Sciences, Virginia Polytechnic Institute and State University, Memoir 2, 41–47.

Burrett, C. (1973): Ordovician biogeography and continental drift. Palaeogeography, Palaeoclimatology, Palaeoecology, 13, 161—201.

Cocks, L.R.M., and Fortey, R.A. (1982): Faunal evidence for oceanic separations in the Palaeozoic of Britain. Journal of the Geological Society of London, 139, 465—478.

Cooper, G.A. (1956): Chazyan and related brachiopods. Smithsonian Miscellaneous Collections, 127, 1—1024.

Courtessole, R. (1973): Le Cambrien moyen de la Montagne Noire. Biostratigraphie. Toulouse. 248 pp.

Courtessole, R. (1975): Un genre de Paradoxididae (Centropleurinae) noveau pour la Montagne Noire (France Méridionale), Bulletin de la Société d'Histoire Naturelle de Toulouse, 111, 211—214.

Cowie, J.W. (1971): Lower Cambrian faunal provinces. Pp. 31—46 in Middlemiss, F.A. and others (eds.), Faunal provinces in space and time. Seel House Press, Liverpool.

Cowie, J.W., Rushton, A.W.A., and Stubblefield, C.J. (1972): A correlation of Cambrian rocks in the British Isles. Geological Society of London, Special Report 1, 40 p.

Curry, G.B., Ingham, J.K., Bluck, B.J., and Williams, A. (1982): The significance of a reliable Ordovician age for some Highland Border rocks in central Scotland. Journal of the Geological Society of London, 139, 451—454.

Dean, W.T. (1960): The use of shelly faunas in a correlation of the Caradoc Series in England, Wales and parts of Scandinavia. Proceedings International Geological Congress, 21st Session, Copenhagen, 7, 82—87.

Dean, W.T. (1966): The Lower Ordovician stratigraphy and trilobites of the Landeyran Valley and the neighbouring district of the Montagne Noire, south-western France. British Museum (natural History), Bulletin, Geology, 12, 245—353.

Dean, W.T. (1967): The distribution of Ordovician shelly faunas in the Tethyan region. Systematics Association Publication, 7, 11—44.

Dean, W.T. (1971): Ordovician trilobites from the Central Volcanic Mobile Belt at New World Island, northeastern Newfoundland. Geological Survey of Canada, Bulletin 210, 37 pp.

Dean, W.T. (1976): Some aspects of Ordovician correlation and trilobite distribution in the Canadian Appalachians. Pp. 227—250 in Bassett, M.G. (ed.), The Ordovician System: proceedings of a Palaeontological Association symposium, Birmingham, September, 1974. University of Wales Press and National Museum of Wales, Cardiff, 696 p.

Dean, W.T. (1979): Trilobites from the Long Point Group (Ordovician), Port au Port Peninsula, southwestern Newfoundland. Geological Survey of Canada, Bulletin, 290, 1—53.

Dean, W.T. (1982): Middle Cambrian trilobites from the Sosink Formation, Derik-Mardin district, south-eastern Turkey. British Museum (natural History), Bulletin, Geology, 36, 1—41.

Dean, W.T., and Martin, F. (1978): Lower Ordovician acritarchs and trilobites from Bell Island, eastern Newfoundland. Geological Survey of Canada, Bulletin, 284, 1—35.

Erben, H.K. (1958): Blinding and extinction of certain Proetidae: (Tril.). Journal of the Palaeontological Society of India, 3, 82—104.

Erdtmann, B.-D. (1982): Palaeobiogeography and environments of planktic dictyonemid graptolites during the earliest Ordovician. Pp. 9—27 in Bassett, M.G. and Dean, W.T. (eds.). The Cambrian-Ordovician boundary: sections, fossil distributions, and correlations. 227 pp. National Museum of Wales, Geological Series No. 3, Cardiff.

Fortey, R.A. (1975): Early Ordovician trilobite communities. Fossils and Strata, 4, 331—352.

Fortey, R.A., Landing, E., and Skevington, D. (1982): Cambrian-Ordovician boundary sections in the Cow Head Group, western Newfoundland. Pp. 95—129 in Bassett, M. and Dean, W.T. (eds.). The Cambrian-Ordovician boundary: sections, fossil distributions, and correlations. 227 pp. National Museum of Wales, Geological Series No. 3, Cardiff.

Fortey, R.A., and Morris, S.F. (1982): The Ordovician trilobite *Neseuretus* from Saudi Arabia, and the palaeogeography of the *Neseuretus* fauna related to Gondwanaland in the earlier Ordovician. British Museum (natural History), Bulletin, Geology, 36, 63—75.

Fortey, R.A., and Owens, R.M. (1978): Early Ordovician (Arenig) stratigraphy and faunas of the Carmarthen district, south-west Wales. British Museum (natural History), Bulletin, Geology, 30, 225—294.

Fritz, W.H. (1972): Lower Cambrian trilobites from the Sekwi Formation type section, Mackenzie Mountains, northwestern Canada. Geological Survey of Canada, Bulletin, 212, 1—90.

Grönwall, G.A. (1902): Bornholms Paradoxideslag og deres Fauna. Denmarks geologiske Unsersøgelse, 13, 1—230.

Hammann, W., Robardet, M., and Romano, M. (1982): The Ordovician System in southwestern Europe (France, Spain, and Portugal). International Union of Geological Sciences, Publication 11, 47 pp.

Harland, W.B., and Gayer, R.A. (1972): The Arctic Caledonides and earlier oceans. Geological Magazine 109, 289–314.

Harrington, H.J., and Leanza, A.F. (1957): Ordovician trilobites of Argentina. University of Kansas (Lawrence), Department of Geology, Special Publication 1, 276 pp.

Hayes, A.D., and Howell, B.F. (1937): Geology of Saint John, New Brunswick. Geological Society of America, Special Paper 5, 1–146.

Henningsmoen, G. (1957): The trilobite family Olenidae with description of Norwegian material and remarks on the olenid and Tremadocian Series. Skrifter utgitt av Det Norske Videnskaps-Akademi i Oslo, I. Mat.-Naturv. Klasse, 1, 303 pp.

Howell, B.F. (1925): Faunas of the Cambrian *Paradoxides* Beds at Manuels, Newfoundland. Bulletins of American Palaeontology, 11 (no. 43), 140 pp.

Hughes, C.P., Ingham, J.K., and Addison, R. (1975): The morphology, classification and evolution of the Trinucleidae (Trilobita). Philosophical Transactions of the Royal Society of London, B, 272, 537–607.

Hupé, P. (1953): Contribution à l'étude du Cambrien inférieur et du Précambrien III de l'Anti-Atlas Marocain. Notes et Mémoires du Service des Mines, Carte géologique du Maroc, 103, 1–402.

Hurst, J.M. (1979): The stratigraphy and brachiopods of the upper part of the type Caradoc of south Salop. British Museum (natural History), Bulletin, Geology, 32, 183–304.

Hutchinson, R.D. (1952): The stratigraphy and trilobite faunas of the Cambrian rocks of Cape Breton Island, Nova Scotia. Geological Survey of Canada, Memoir 263, 124 p.

Hutchinson, R.D. (1962): Cambrian stratigraphy and trilobite faunas of southeastern Newfoundland. Geological Survey of Canada, Bulletin 88, 148 p.

Jaanusson, V. (1972): Aspects of carbonate sedimentation in the Ordovician of Baltoscandia. Lethaia, 6, 11–34.

Kay, M., and Eldredge, N. (1968): Cambrian trilobites in central Newfoundland volcanic belt. Geological Magazine, 105, 372–377.

Kokelaar, B.P. (1979): Tremadoc to Llanvirn volcanism on the southeast side of the Harlech dome (Rhobell Fawr), N. Wales. Pp. 591–596 in Harris, A.L., Holland, C.H., Leake, B.E. (eds.), The Caledonides of the British Isles – reviewed. Geological Society of London, 768 p.

Lister, T.R., Cocks, L.R.M., and Rushton, A.W.A. (1969): The basement beds in the Bobbing borehole, Kent. Geological Magazine, 106, 601–603.

Lochman-Balk, C., and Wilson, J.L. (1958): Cambrian biostratigraphy in North America. Journal of Palaeontology, 32, 312–350.

Lu, Y-H. and others (1965): Fossils of each group of China: Chinese trilobites. (in Chinese). 766 pp. Peking.

Martin, F. (1982): Some aspects of late Cambrian and early Ordovician acritarchs. Pp. 29–40 in Bassett, M.G., Dean, W.T. (eds.). The Cambrian-Ordivician boundary: sections, fossil distributions, and correlations. National Museum of Wales, Geological Series, 3.

Martinsson, A. (1974): The Cambrian of Norden. Pp. 185–284 in Holland, C.H. (ed.). Cambrian of the British Isles. John Wiley and Sons (New York).

McKerrow, W.S., and Cocks, L.R.M. (1976): Progressive faunal migration across the Iapetus Ocean. Nature, 263, 304–305.

Nelson, C.A., and Hupé, P. (1964): Sur l'existence de Fallotaspis et Daguinaspis, Trilobites marocains, dans le Cambrien inférieur de Californie et ses conséquences. Compte Rendu Hebdomadaire des Séances de l'Academie des Sciences, Paris, 258, 621–623.

Neuman, R.B. (1964): Fossils in Ordovician tuffs, northeastern Maine. U.S. Geological Survey Bulletin 1181-E, 38 p.

Neuman, R.B. (1968): Paleogeographic implications of Ordovician shelly fossils in the Magog Belt of the northern Appalachian region. Pp. 35–48 in Zen, E. and others (eds.). Studies of Appalachian geology: northern and maritime. xviii + 475 pp. Interscience, New York.

Neuman, R.B. (1972): Brachiopods of early Ordovician volcanic islands. Proceedings of the 24th International Geological Congress, Section 7, 297–302.

Neuman, R.B., and Bates, D.E.B. (1978): Reassessment of Arenig and Llanvirn age (early Ordovician) brachiopods from Anglesey, north-west Wales. Palaeontology, 21, 571–613.

Neuman, R.B., and Bruton, D.L. (1974): Early middle Ordovician fossils from the Hølonda area, Trondheim region, Norway. Norsk Geologisk Tidsskrift, 54, 69–115.

Norford, B.S. (1969): The early Canadian (Tremadocian) trilobites *Clelandia* and *Jujuyaspis* from the southern Rocky Mountains of Canada. Geological Survey of Canada, Bulletin 1982, 1–15.

Öpik, A.A. (1961): Alimentary caeca of agnostids and other trilobites. Palaeontology, 3, 410–438.

Owens, R.M., Fortey, R.A., Cope, J.C.W., Rushton, A.W.A. and Bassett, M.G. (1982): Tremadoc faunas from the Carmarthen district, South Wales. Geological Magazine, 119, 1–38.

Palmer, A.R. (1969): Cambrian trilobite distributions in North America and their bearing on Cambrian paleogeography of Newfoundland. American Association of Petroleum Geologists, Memoir 12, 139−144.
Palmer, A.R. (1971): The Cambrian of the Appalachian and eastern New England regions, eastern United States. Pp. 169−217 in Holland, C.H. (ed.). Cambrian of the New World. John Wiley and Sons.
Palmer, A.R. (1972): Problems of Cambrian biogeography. Proceedings of the 24th International Geological Congress, Section 7, 310−315.
Palmer, A.R. (1973): Cambrian trilobites. Pp. 3−11 in Hallam, A. (ed.), Atlas of Palaeobiogeography. Elsevier Scientific Publishing Company.
Palmer, A.R. (1979): Cambrian. In Robison, R.A. and Teichert, C. (eds.). Treatise on Invertebrate Palaeontology. Part A. Introduction, A119−A135. Geological Society of America and University of Kansas Press.
Palmer, A.R., and Stewart, J.H. (1968): A paradoxidid trilobite from Nevada. Journal of Paleontology, 42, 177−179.
Robison, R.A. (1972): Mode of life of agnostid trilobites. Report of the 24th Session International Geological Congress, Montreal, Section 7, 33−40.
Robison, R.A., and Pantoja-Alor, J. (1968): Tremadocian trilobites from the Nochixtlán region, Oaxaca, Mexico. Journal of Paleontology, 42, 767−800.
Rodgers, J. (1965): Long Point and Clam Bank formations, western Newfoundland. Geological Association of Canada, Proceedings, 16, 83−94.
Rose, E.R. (1952): Torbay map-area, Newfoundland. Geological Survey of Canada, Memoir, 265, 1−64.
Ross, R.J. (1949): Stratigraphy and trilobite faunal zones of the Garden City Formation, northeastern Utah. American Journal of Science, 247, 472−491.
Ross, R.J. (1967): Calymenid and other Ordovician trilobites from Kentucky and Ohio. United States Geological Survey, Professional Paper 583-B, B1−B19.
Ross, R.J. (1975): Early Paleozoic trilobites, sedimentary facies, lithospheric plates, and ocean currents. Fossils and Strata, 4, 307−329.
Ross, R.J., and Ingham, J.K. (1970): Distribution of the Toquima-Table Head (Middle Ordovician Whiterock) Faunal Realm in the northern hemisphere. Geological Society of America, Bulletin, 81, 393−408.
Ross, R.J. and others (1982): The Ordovician System in the United States. International Union of Geological Sciences, Publication 12, 73 pp.
Rushton, A.W.A. (1966): The Cambrian trilobites from the Purley Shales of Warwickshire. Palaeontological Society [Monograph], 120, 55 pp.
Rushton, A.W.A. (1982): The biostratigraphy and correlation of the Merioneth-Tremadoc Series boundary in North Wales. Pp. 41−59 in Bassett, M.G. and Dean, W.T. (eds.). The Cambrian-Ordovician boundary: sections, fossil distributions, and correlations. National Museum of Wales, Geological Series No. 3, Cardiff.
Rushton, A.W.A., and Tripp, R.P. (1979): A fossiliferous lower Canadian (Tremadoc) boulder from the Benan Conglomerate of the Girvan district. Scottish Journal of Geology, 15, 321−327.
Schenk, P.E. (1971): Southeastern Atlantic Canada, northwestern Africa, and continental drift. Canadian Journal of Earth Sciences, 8, 1218−1251.
Schuchert, C., and Cooper, G.A. (1930): Upper Ordovician and Lower Devonian stratigraphy and paleontology of Percé, Quebec. P.I. C. Schuchert. Stratigraphy and faunas. American Journal of Science, (5) 20, 161−176. P.II. G.A. Cooper. New species from the Upper Ordovician of Percé. Ibid., 265−288, 365−392.
Sdzuy, K. (1971): La subdivision bioestratigrafica y la correlacion del Cambrico Medio de España. Publ. I. Congr. Hisp. Luso Amer. Geol. Econ., 2, Seccion 1, 769−782.
Sdzuy, K. (1972): Das Kambrium der Acadobaltischen Faunenprovinz. Zentralblatt für Geologie und Paläontologie, II, 1−91.
Sdzuy, K. (1978): The Precambrian-Cambrian boundary beds in Morocco (Preliminary Report). Geological Magazine, 15, 83−94.
Seilacher, A. (1963): Kaledonischer Unterbau der Irakiden. Neues Jahrbuch für Geologie und Paläontologie, Monatshefte, 10, 527−542.
Shaw, A.B. (1951): Paleontology of Northwestern Vermont. I. New Late Cambrian trilobites. Journal of Paleontology, 25, 97−114.
Shaw, F.C., and Fortey, R.A. (1977): Middle Ordovician facies and trilobite faunas in N. America. Geological Magazine, 114, 409−443.

Sheehan, P.M. (1975): Brachiopod synecology in a time of crisis (Late Ordovician-Early Silurian). Paleobiology, 1, 205—212.
Skinner, R. (1974): Tetagouche Lakes, Bathurst, and Nepisiguit Falls map-areas. Geological Survey of Canada, Memoir, 371, 1—133.
Smith, A.G., Hurley, A.M., and Briden, J.C. (1981): Phanerozoic paleocontinental world maps. Cambridge University Press. 102 p.
Spjeldnaes, N. (1961): Ordovician climatic zones. Norsk Geologisk Tidsskrift, 41, 45—77.
Spjeldnaes, N. (1981): Lower Palaeozoic palaeoclimatology. Pp. 199—256 in Holland, C.H. (ed.), Lower Palaeozoic of the Middle East, eastern and southern Africa, and Antarctica. John Wiley and Sons. Ltd.
Størmer, L. (1940): Early descriptions of Norwegian trilobites. The type specimens of C. Boeck, M. Sars and M. Esmark. Norsk Geologisk Tidsskrift, 20, 113—151.
Størmer, L. (1967): Some aspects of the Caledonian geosyncline and foreland west of the Baltic Shield. Quarterly Journal of the Geological Society of London, 123, 183—214.
Strömberg, A.G.B. (1981): The European Caledonides and the Tornquist Lineament. Geologiska Föreningens i Stockholm Förhandlingar, 103, 167—171.
Stubblefield, C.J., and Bulman, O.M.B. (1927): The Shineton Shales of the Wrekin district: with notes on their development in other parts of Shropshire and Herefordshire. Quarterly Journal of the Geological Society of London, 83, 55—146.
Van Ingen, G. (1914): Cambrian and Ordovician faunas of southeastern Newfoundland. Geological Society of America, Bulletin, 25, 138.
Westergård, A.H. (1946): Agnostidea of the Middle Cambrian of Sweden. Sveriges Geologiska Undersökning, C, 477, 1—141.
Westergård, A.H. (1949): Non-agnostidean trilobites of the Middle Cambrian of Sweden. II. Sveriges Geologiska Undersökning, C, 511, 1—57.
Whitcomb, L. (1930): New information on *Homalonotus trentonensis*. Bulletin of the Geological Society of America, 41, 341—350.
Whittard, W.F. (1966): The Ordovician trilobites of the Shelve Inlier, west Shropshire, Pt. VIII. Palaeontographical Society Monograph, 265—306.
Whittington, H.B. (1953): A new Ordovician trilobite from Florida. Breviora, Harvard University Museum of Comparative Zoology, 17, 1—6.
Whittington, H.B. (1959): Silicified Middle Ordovician trilobites. Remopleurididae, Trinucleidae, Raphiophoridae, Endymioniidae. Bulletin of the Museum of Comparative Zoology at Harvard College, 121, 369—496.
Whittington, H.B. (1963): Middle Ordovician trilobites from Lower Head, western Newfoundland. Bulletin of the Museum of Comparative Zoology, Harvard College, 129 (1), 1—118.
Whittington, H.B. (1965): Trilobites of the Ordovician Table Head Formation, western Newfoundland. Bulletin of the Museum of Comparative Zoology, Harvard University, 132 (4), 275—442.
Whittington, H.B. (1966): Phylogeny and distribution of Ordovician trilobites. Journal of Paleontology, 40, 696—737.
Whittington, H.B., and Hughes, C.P. (1972): Ordovician geography and faunal provinces deduced from trilobite distribution. Philosophical Transactions of the Royal Society of London, B 262, 235—278.
Whittington, H.B., and Hughes, C.P. (1974): Geography and faunal provinces in the Tremadoc epoch. Pp. 203—218 in Ross, C.A. (ed.). Palaeogeographic provinces and provinciality. Society of Economic Paleontologists and Mineralogists, Special Publication 21, 203—218.
Whittington, H.B., and Kindle, C.H. (1969): Cambrian and Ordovician stratigraphy of western Newfoundland. American Association of Petroleum Geologists, Memoir 12, 655—664.
Whittington, H.B., and Orchard, M.J. (1977): Upper Ordovician fauna from northwest Iran. Third International Symposium on the Ordovician System, Columbus 1977, Abstracts, p. 32.
Williams, A. (1972): An Ordovician Whiterock fauna in Western Ireland. Proceedings of the Royal Irish Academy, 72, B, 209—219.
Williams, A. (1973): Distribution of brachiopod assemblages in relation to Ordovician palaeogeography. Special Papers in Palaeontology, 12, 241—269.
Williams, A. (1976): Plate tectonics and biofacies evolution as factors in Ordovician correlation. Pp. 29—66 in Bassett, M.G. (ed.). The Ordovician System; proceedings of a Palaeontological Association symposium, Birmingham, September, 1974. University of Wales Press and National Museum of Wales, Cardiff.
Williams, A. and others (1972): A correlation of Ordovician rocks in the British Isles. Geological Society of London, Special Report 3, 74 pp.

Wilson, J.L. (1957): Geography of olenid trilobite distribution and its influence on Cambro-Ordovician correlation. American Journal of Science, 255, 321–340.
Wilson, J.T. (1966): Did the Atlantic close and then re-open? Nature, 211, 676–681.
Windley, B.F. (1977): The evolving continents. Wiley and Sons, London. 385 pp.
Winston, D., and Nicholls, H. (1967): Late Cambrian and Early Ordovician faunas from the Wilberns Formation of central Texas. Journal of Paleontology, 41, 66–96.
Wright, A.D. (1968): A westward extension of the upper Ashgillian *Hirnantia* fauna. Lethaia, 1, 352–367.
Ziegler, A.M., Hansen, K.S., Johnson, M.E., Kelly, M.A., Scotese, C.R., and Van der Voo, R. (1977): Silurian continental distributions, palaeogeography, climatology, and biogeography. In McElhinny, M.W. (ed.), The Past Distribution of Continents. Tectonophysics, 40, 13–51.

Revised manuscript received 15 March 1984

The North American Appalachian Orogen

Robert D. Hatcher, Jr.
Department of Geology, University of South Carolina, Columbia, SC 29208, USA

Key Words

Avalon Terrane
Appalachian Orogen
Taconic Orogen
(Acadian Orogen)
Alleghanian Orogen

Abstract

The Appalachian orogen as presently exposed, extends from Newfoundland to Alabama in eastern North America and consists of a Palaeozoic orogen developed along the eastern Precambrian continental margin of North America. Its history records considerable diversity in stratigraphic, deformational, metamorphic, and plutonic events. Following the Proterozoic (1 GA) Grenville orogeny in eastern North America, a passive continental margin developed in the late Precambrian and began to receive continental rise, slope, and shelf sediments from the eroding Grenville Mountains along with associated rift felsic and mafic volcanic rocks. The Avalon terrane consists of sequences of late Precambrian to early Palaeozoic volcanic rocks and clastic to volcaniclastic sediments that appear unrelated to the North American margin at the time the platform sequence was being deposited along the late Precambrian coastline. This zone is represented by the Carolina slate belt in the eastern Piedmont of the southern Appalachians and in the Narrangansett Basin-Boston platform area of southeastern New England. In Middle Ordovician time, the eastern part of the margin began to subside rapidly in response to loading of a part of the crust farther east by advancing thrust sheets of the Taconic orogen. Large sediment accumulations to the west reflect uplift taking place in the central part of the orogen doubtlessly related to contemporaneous thrusting and penetrative deformation, Barrovian metamorphism (high pressure-low temperature in Vermont), and plutonism. Basement rocks belonging to the Grenville cycle were incorporated into the deforming metamorphic core during the Taconic orogeny. These were remobilized and generally thrust toward the craton. During Devonian time, the rocks of central and southern New England were subjected to high-temperature relatively high-pressure metamorphism, and many of the earlier plutons and older structures were reactivated and to some degree overprinted by structures representing the Acadian event. It is uncertain to what degree Acadian metamorphism affected the crystalline rocks of the Blue Ridge and Piedmont in the southern Appalachians. Alleghanian events in North America include the deformation of the foreland; deposition of molasse-type sediments on the craton west of and within the Appalachian orogen; deformation and metamorphism along the eastern margin of the Appalachians in the southern Appalachian Piedmont, southeastern New England, and possibly along the coast of the Maritime Appalachians in Canada; and the intrusion of plutons, mostly granites, in association with the deformational-metamorphic events. There is a possibility that large segments of the Appalachian orogen have been transported along strike for tens or even hundreds of kilometers to be emplaced in their present positions within the orogen. The only part of the Appalachian orogen that can be considered reasonably intact and relatable to the North American carton is the late Precambrian-Palaeozoic continental margin. The internal parts were developed offshore and may well have been moved for considerable distances along strike prior to the final closing and collision events that terminated orogenic activity throughout the mountain chain. Even so, most agree that large parts of the Appalachian orogen, particularly in the southern half, have been overthrust onto the old continental margin.

Introduction

The North American Appalachian orogen consists of a Palaeozoic orogen developed along the eastern Precambrian continental margin of North America. Its history records considerable diversity in stratigraphic, deformational, metamorphic, and plutonic events. My purpose here is to outline the various subdivisions of the orogen and comment upon what we know of its history at the present time.
The Appalachian orogen, as presently exposed, extends from Newfoundland to Alabama in eastern North America (Fig. 1). The subdivision outlined below for the Appalachians represent a time-space categorization that may serve to break down the classic geographic subdivisions that have historically dominated our descriptions of the Appalachians. These subdivisions provide a means by which the history of the orogen may be more easily described and related to the constructive processes that built it on the eastern late Precambrian margin of North America during Palaeozoic time. These subdivisions reflect the likely plate tectonics history of the orogen and the manner in which the present southeastern continental margin of North America evolved throughout Palaeozoic time.

Subdivisions of the Appalachian Orogen

Late Precambrian — Early Palaeozoic Passive Margin

Following the Proterozoic (1 GA) Grenville orogeny in eastern North America, a passive continental margin developed in the late Precambrian and began to receive sediments from the eroding Grenville terrane. These sediments constituted a continental rise, slope and shelf sequence that evolved in time into a stable trailing margin sequence. Sediment thicknesses range from zero to twelve to fifteen thousand meters. In addition, extensive associated felsic and mafic volcanic rocks occur in several areas throughout the southern and central Appalachians (Rankin, 1976). Initially, in part of the southern Appalachians, the sediments were relatively mature. Subsequently, a great thickness of poorly sorted sediments appeared, reaching its greatest thickness as the Great Smoky Group in the southern Appalachians. Later, the character of sediments changed again to a more mature assemblage before the beginning of Cambrian time.

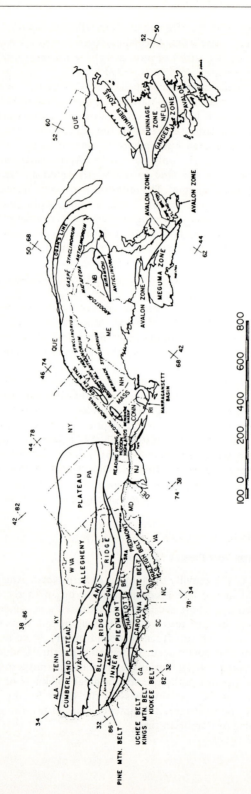

Fig. 1
Map showing the major subdivisions of the North American Appalachian orogen. AA — Alto allochthon. GMW — Grandfather Mountain window. SRA — Smith River Allochthon.

The basal Cambrian clastic succession in the Appalachians is more uniform and dominantly more mature than the underlying succession of Precambrian sedimentary rocks. Overlying the basal Cambrian clastic sequence is a succession of carbonates, then clastic and more carbonates, that comprise the stable margin carbonate bank and associated sediments that existed in eastern North America from Early Cambrian time until the end of Early Ordovician time. The clastics that occur in this sequence are derived entirely from the craton. The Cambrian (?) Unicoi clastic-mafic volcanic assemblage that occurs in the central and southern Appalachians is the last vestige of volcanism in the evolving continental-margin sequence prior to Middle Ordovician time.

Avalon Terrane

The Avalon Terrane (Williams, 1978; Williams and Hatcher, 1983) consists of sequences of late Precambrian to early Palaeozoic volcanic rocks and clastic to volcaniclastic sediments that appear unrelated to the North American margin at the time the platform sequence was being deposited along the late Precambrian coastline. This zone is represented by the Carolina slate belt in the eastern Piedmont of the southern Appalachians and in the Narrangansett Basin-Boston platform area of southeastern New England. The Carolina slate belt in the South has been suggested to be sufficiently different by Secor and others (1983) to be termed the Carolina terrane as a subdivision of the Avalon terrane. A part of the coastal Maine carbonates and metaclastics probably also represent the Avalon terrane. High grade equivalents of the Carolina slate belt rocks probably comprise most of the country rocks of the Charlotte and Kiokee belts and part of the Raleigh belt in the southern Appalachians (Snoke and others, 1980). There are no outcrops of rocks assignable to the Avalon terrane in the central Appalachians. They are either covered by the Atlantic Coastal Plain or were removed or not deposited in this sector. Volcanic rocks in the Avalon terrane exhibit a variety of compositions ranging from mafic to felsic. They have been variously interpreted as a system of volcanic rocks belonging to an island arc, a rifted margin, or a sedimentary-volcanic arc sequence deposited on continental crust (Hatcher, 1972). Gravity data suggest that the rocks of the Avalon terrane are underlain by a mafic or intermediate crust because the gravity field beneath the southern Appalachians part of the zone is high (Hatcher and Zietz, 1980).

Taconic Orogen

In Middle Ordovician time, following regional uplift of eastern North America and development of an extensive unconformity on top of the marginal carbonate bank, shallow seas encroached upon the continent once more and the carbonate bank was briefly reestablished. Shortly thereafter, however, the eastern part of the margin began to subside rapidly forming a foredeep basin in which black shale was deposited. This rapid subsidence probably represented loading of a part of the crust farther east by advancing thrust sheets of the Taconic orogen. In New England, the Taconic klippes, preceded by debris, were emplaced into the basin from the east with concomitant ophiolite emplacement farther north. Farther south in Pennsylvania, the Hamburg klippe was emplaced, and large amounts of clastic sediments were deposited from Middle Ordovician until Early Silurian time in the southern and central Appalachians. Accumulations reached a maximum thickness of a few thousand meters in areas of east Tennessee and central Pennsylvania. These reflect uplift taking place in the central part of the orogen probably related to contemporaneous thrusting and penetrative deformation, Barrovian metamorphism (high pressure-low temperature in Vermont), and plutonism. Plutonic activity in the form of granitic, intermediate, and mafic plutons is found throughout the Appalachian orogen in the New England region (Robinson and Hall, 1980) and occurs over wide areas, but primarily in the central part of the orogen in the southern and central Appalachians. In northern Maine there is an Arenigian unconformity that probably represents the oldest response to Taconic activity in the internal parts of the orogen. An ocean of unknown size was closed at this time resulting in the emplacement of ultramafic bodies along a prominent trend in the western part of the orogen east of the primary continental margin sediments and basement rocks of the craton, that is, in the central to eastern Blue Ridge of the southern Appalachians, the Piedmont of the central Appalachians, and east of the basement massifs in New England. Relict blueschist metamorphism occurs in north-central Vermont (Laird and Albee, 1981).

Basement rocks belonging to the Grenville cycle were incorporated into the deforming metamorphic core during the Taconic orogeny. These were remobilized, sometimes detached, penetratively deformed to a greater or lesser degree, and generally thrust toward the craton. Many were reactivated in the Acadian orogeny and moved farther westward on large faults (Hatcher, 1978, 1981, 1983).

Although the Taconic event has been considered to be a relatively minor event in the central and eastern New England Appalachians, it is a major event in western New England (Robinson and Hall, 1980). It is also considered to be the major thermal event affecting the rocks of the Blue Ridge and western Piedmont, as well as parts of the eastern Piedmont farther south (Butler, 1972). However, the evidence concerning the role of later orogenies and the degree to which they affected the southern Appalachians is presently inconclusive, although more evidence favouring major Acadian and Alleghanian metamorphism and plutonism is appearing.

Acadian Orogen

During Silurian and Early Devonian times, there was a return to relatively quiet continental margin or supracrustal conditions over most of the Appalachians. In the continental interior, the Silurian-Devonian rocks are represented by carbonates and relatively clean clastics. However, volcanic and deep-water sedimentary rocks are found in the Silurian and Devonian of New England, possibly indicating a return of continental margin or deep basin conditions. Some of the sediments formed at this time could be the successors to those of the earlier continental margin that was overridden during the Taconic event. Some of the Silurian volcanic rocks were formed under subaerial conditions.

During Devonian time, the rocks of central and southern New England were subjected to high-temperature relatively high-pressure metamorphism, and many of the earlier plutons and older structures were reactivated and to some degree overprinted by structures representing this Acadian event. Abundant plutons are associated with this event in New England and most are S-type granites (Robinson and Hall, 1980). It is uncertain to what degree Acadian metamorphism affected the crystalline rocks of the Blue Ridge and Piedmont in the southern Appalachians. It has been suggested that the Acadian event was an event of major faulting, based upon Rb/Sr whole-rock ages obtained from dating mylonites (Odom and Fullagar, 1973) and may be the major event in the Piedmont (Dallmeyer, 1978; Odom and others, 1982).

South of Virginia, Silurian and Devonian sediments are practically nonexistent on the craton in southern Kentucky, Tennessee, and north Georgia. Farther south there are some sediments of this age in Alabama, but for most areas of the southern Appalachians, the Devonian and Silurian on the foreland were times of little deposition. There is no evidence in the South of a Devonian clastic wedge that might reflect the Acadian orogeny. The Talladega sequence of Alabama and Georgia must be a pre-Acadian clastic-volcanic sequence. It contains Lower Devonian fossils and was affected by apparent Acadian dynamothermal events (Tull, 1978).

Alleghanian Orogen

The Alleghanian is a Late Carboniferous-Permian event roughly correlative to the Variscan-Hercynian events of Western Europe and northwestern Africa. Associated with the Alleghanian events in North America are the deformation of the foreland; deposition of molasse-type sediments on the craton west of and within the Appalachian orogen; deformation and metamorphism along the eastern margin of the Appalachians in the southern Appalachian Piedmont, southeastern New England, and possibly along the coast of the Maritime Appalachians in Canada; and the intrusion of plutons, mostly granites, in association with the deformational-metamorphic events.

Alleghanian deformation in the Appalachian foreland postdates the extensive clastic deposition in the so-called Appalachian basin from Alabama to Pennsylvania. It is characterized by thin-skinned thrusting directed toward the continent in which thrust sheets are detached from the underlying basement and follow incompetent zones in the stratigraphic section. Toward the west they ramp across the more competent zones to higher incompetent detachments. Deformation involves rocks as young as early Permian in West Virginia, Pennsylvania, and Ohio and as young as Pennsylvanian in the Valley and Ridge of Pennsylvania and several places farther south. Thus, the timing of this event in the foreland of the Appalachians is reasonably well established. Thrusts of the Valley and Ridge are overridden in the southern Appalachians by the Blue Ridge thrust sheet, which along its toe in east Tennessee and Georgia involves Mississippian rocks. These rocks serve to indicate that movement on the Blue Ridge thrust is likely Alleghanian throughout most of its extent. Farther to the southeast, vestiges of movement during the Alleghanian are more difficult to resolve within the crystalline complex. It is likely that brittle deformation along the Brevard fault which brings up slices of carbonate from beneath the thrust sheet is Alleghanian (Hatcher, 1978). The metamorphism and penetrative deformation occurring in the Kiokee and Raleigh belts of the eastern Piedmont have been dated as Alleghanian by cross-cutting relationships of deformed plutons. The Modoc and Augusta faults are likewise Alleghanian structures, since they are intimately involved in the deformation occurring in this area (Snoke and others, 1980). In the New England area there is some suggestion on the foreland of Alleghanian deformation in northeastern New York and Vermont (Geiser and Engelder, 1983). However, this is not well established. Geiser and Engelder (1983) have also presented evidence for two pulses in the Alleghanian. In southeastern New England, metamorphism, plutonic activity, and penetrative deformation associated with deformation of the Pennsylvanian rocks of the Narrangansett basin are well established. Mosher (1983) has suggested that these may be rhomb grabens related to large strike-slip faults. Some of the large brittle faults that occur in southeastern Maine and extend into the Canadian Maritime region may likewise be Alleghanian features.

Appalachian-Ouachita Connections

The nature of the connections between the Ouachita and Appalachian orogens has long been a subject of discussion and concern (Thomas 1976, 1983). No precise answer to the question is yet available because the area where the connection occurs is buried beneath the Cretaceous and Tertiary sediments of the Gulf Coastal Plain in the Mississippi Embayment. Subsurface and geophysical studies indicate that a belt of deformed rocks extends between the exposed Appalachian and Ouachita structures. New gravity and magnetic studies (Zietz and others, 1982; Higgins and Zietz, 1983) show that geophysical trends reflecting basement rocks are truncated on the southwest against the Ouachita structural front. Major differences in contemporaneous facies and thicknesses are demonstrated by stratigraphic sections exposed in the Appalachians and Ouachitas in Alabama and Arkansas. The stratigraphic transition is documented by subsurface data (Thomas, 1976).

Constructive History

The Appalachian orogen was built along the Precambrian continental margin of North America following the rifting and opening of an ancient ocean. The size of this ocean is presently indeterminant. The evidence confirming that the continents were rifted to produce this ocean is present in the late Precambrian-Cambrian sedimentary and volcanic rocks of eastern North America. Following the rifting and spreading events in late Precambrian time, the eastern margin of North America developed into a carbonate bank platform that existed until the end of Early Ordovician time. At this time the platform was uplifted, eroded, and subsided once more. Then a foredeep basin formed along the eastern flank of the North American continent prior to influx of clastic sediments. The arrival of the first thrust sheets onto this flank apparently occurred in Ordovician time. These carried fragments of oceanic crust and mantle as ophiolites shedding debris. Whether the entire ocean basin or only a part of it was closed at this time remains to be determined, but a segment of oceanic crust was destroyed. Accompanying the closing of this part of the ocean were pulses of extensive penetrative deformation, metamorphism, and plutonic activity from Early to Middle Ordovician until Early Silurian times. These pulses constitute the Taconic orogeny.
Sedimentation, which was limited in the South but more extensive in the North, followed a period of quiesence and was accompanied by some volcanic activity. This was followed by another compressional event, perhaps related to the closing of an ocean in the North. It produced high-temperature metamorphism in New England, faulting and possibly metamorphism in the southern Appalachians. Extensive emplacement of plutons of various compositions occurred throughout the Appalachians at this time. These events constitute the Acadian orogeny.
The Alleghanian orogeny occurred near the end of Palaeozoic time and terminated compressive tectonic activity in the Appalachian orogen. This event may be related to the final closing of the ancient large ocean that may have existed off eastern North America during Palaeozoic time. The discrepancy in stratigraphic, deformational, metamorphic, and other evidence along the orogen indicates that perhaps the collision between New England and another continent may have involved Africa or South America, or possibly both. This would help account for the diachronous nature of the collisions and the apparent discrepancy in the nature of the deformations that occur up and down the orogen.
The variable that must be considered, in addition to across-strike transport, in any reconstructive hypothesis for the Appalachian orogen is the possibility that large segments of the orogen have been transported along strike for tens or even hundreds of kilometers to be emplaced in their present positions within the orogen (Williams and Hatcher, 1983). The only part of the Appalachian orogen that can be considered reasonably intact and

relatable to the North American craton is the late Precambrian-Palaeozoic continental margin. The internal parts were developed offshore and may well have been moved for considerable distances along strike prior to the final closing and collision events that terminated orogenic activity throughout the mountain chain. Therefore, it is necessary to consider that a number of segments of the Appalachian orogen may be suspect terranes similar in characteristics described by Coney and others (1980) in the North American Cordillera.

Deformation in the foreland is presently thought to be related to transmittal of stresses through the deformed crystalline mass with only a local thermal high on the east side of the deformed mass, resulting in the thin-skinned deformation of the foreland and transportation of some of the crystalline rocks over the foreland along the western edge of the orogen. Emplacement of the Valley and Ridge thrusts by gravity has also been suggested (Gwinn, 1964; Milici, 1975), but this mechanism encounters significant difficulties where problems of the transition from the external foreland parts of the orogen to the internal parts must be considered. Gravity may have played a part in the readjustment process of final emplacement.

Most Appalachian geologists agree that parts of the Appalachian orogen are allochthonous. The distance from a major gravity gradient to the outermost thrusts and the possible amount of overthrusting decrease northward (Price and Hatcher, 1983). Seismic reflection data have led to the proposal that the entire southern sector of the Appalachian orogen was thrust onto the continental margin along a master decollement which was active throughout most of Palaeozoic time (Harris and Bayer, 1979; Cook and others, 1979). The idea that major parts of the crystalline core were thrust over the foreland was suggested earlier based upon occurrences of lower grade carbonate slices in the Brevard zone (Hatcher, 1971, 1978). Magnetic and gravity data, in conjunction with surface geology, also suggest that all of the southern Blue Ridge and much of the Piedmont are allochthonous (Hatcher and Zietz, 1980).

Acknowledgments

An early draft of this chapter benefited from comments by D.R. Wones, Peter Robinson, P.H. Osberg, D.W. Rankin, W.A. Thomas, and A.R. Palmer.

References

Butler, J.R. (1972): Age of Palaeozoic regional metamorphism in the Carolinas, Georgia and Tennessee southern Appalachians: American Journal of Science, v. 272, p. 319–333.
Coney, P.J., Jones, D.L., Monger, J.W.H. (1980): Cordilleran suspect terrains: Nature, v. 288, p. 329–333.
Cook, F.A., Albaugh, D.S., Brown, L.D., Kaufman, Sidney, Oliver, J.E., and Hatcher, R.D., Jr. (1979): Thin-skinned tectonics in the crystalline southern Appalachians: COCORP seismic-relction profiling of the Blue Ridge and Piedmont: Geology, v. 7, p. 563–567.
Dallmeyer, R.D. (1978): $^{40}Ar/^{39}Ar$ incremental-release ages of hornblende and biotite across the Georgia Inner Piedmont: Their bearing on late Palaeozoic-early Mesozoic tectonothermal history: American Journal of Science, v. 278, p. 124–149.
Geiser, Peter and Engelder, Terry (1983): The distribution of layer parallel shortening fabrics in the Appalachian foreland of New York and Pennsylvania: Evidence for two non-coaxial phases of the Alleghanian orogeny, in, Hatcher, R.D., Jr., Williams, Harold and Zietz, Isidore, eds., Contributions to the Tectonics and Geophysics of Mountain Chains: Geological Society of America Memoir 158, p. 161–176.
Gwinn, V.E. (1964): Thin-skinned tectonics in the Plateau and northwestern Valley and Ridge provinces of the Central Appalachians: Geological Society of America Bulletin, v. 75, p. 863–900.

Harris, L.D. and Bayer, K.C. (1979): Sequential development of the Appalachian orogen above a master decollement — A hypothesis: Geology, v. 7, p. 568—572.
Hatcher, R.D., Jr. (1971): Structural, petrologic and stratigraphic evidence favoring a thrust solution to the Brevard problem: American Journal of Science, v. 270, p. 177—202.
Hatcher, R.D., Jr. (1972): Development model for the southern Appalachians: Geological Society of America Bulletin, v. 83, p. 2735—2760.
Hatcher, R.D., Jr. (1978): Tectonics of the western Piedmont and Blue Ridge southern Appalachians: Review and speculation: American Journal of Science, v. 278, p. 276—304.
Hatcher, R.D., Jr. (1981): Thrusts and nappes in the North American Appalachian Orogen, in, McClay, K.R., and Price, N.J. (eds.), Thrust and Nappe Tectonics: Geological Society of London Special Publication 9, p. 491—499.
Hatcher, R.D., Jr. (1983): Basement massifs in the Appalachians: Their role in deformation during the Appalachian orogenies: Geological Journal, v. 18, p. 255—265.
Hatcher, R.D., Jr. and Zietz, Isidore (1980): Tectonic implications of regional aeromagnetic and gravity data from the southern Appalachians, in, Wones, D.R. (ed.), The Caledonides in the USA: IGCP Project 27, Virginia Polytechnic Institute, Memoir 2, p. 235—244.
Higgins, M.W. and Zietz, Isidore (1983): Geological interpretation of geophysical maps of the pre-Cretaceous 'basement' beneath the Coastal Plain of the Southeastern United States, in, Hatcher, R.D., Jr., Williams, Harold and Zietz, Isidore (eds.), Contributions to the Tectonics and Geophysics of Mountain Chains: Geological Society of America Memoir 158, p. 125—130.
Laird, J. and Albee, A.L. (1981): High pressure metamorphism in mafic schist from northern Vermont: American Journal of Science, v. 281, p. 97—126.
Milici, R.C. (1975): Structural patterns in the southern Appalachians: Evidence for a gravity slide mechanism for Alleghanian deformation: Geological Society of America Bulletin, v. 86, pp. 1316—1320.
Mosher, Sharon (1983): Kinematic history of the Narragansett Basin, Machusetts and Rhode Island: Constraints on late Palaeozoic plate reconstructions: Tectonics, v. 2, p. 327—344.
Odom, A.L. and Fullagar, P.D. (1973): Geochronological and tectonic relationships between the Inner Piedmont, Brevard zone and Blue Ridge belts, North Carolina: American Journal of Science, v. 273-A, p. 133—149.
Odom, A.L., Hatcher, R.D., Jr., and Hooper, R.J. (1982): A pre-metamorphic tectonic boundary between contrasting Appalachian basements, southern Georgia Piedmont: Geological Society of America Abstracts with Programs, v. 14, p. 579.
Price, R.A. and Hatcher, R.D., Jr. (1983): Tectonic significance of similarities in the evolution of the Alabama-Pennsylvania Appalachians and the Alberta-British Columbia Canadian Cordillera, in, Hatcher, R.D., Jr., Williams, Harold and Zietz, Isidore (eds.), Contribution to the Tectonics and Geophysics of Mountain Chains: Geological Society of America, p. 149—160.
Rankin, D.W. (1976): Appalachian salients and recesses: Late Precambrian continental breakup and the opening of the Iapetus Ocean: Journal of Geophysical Research, v. 81, p. 5606—5619.
Robinson, Peter and Hall, L.M. (1980): Tectonic synthesis of southern New England, in, Wones, D.R., (ed.), Caledonides in the U.S.A., IGCP Project 27, Virginia Polytechnic Institute Memoir 2, p. 73—82.
Secor, D.T., Jr., Samson, S.L., Snoke, A.W. and Palmer, A.R. (1983): Confirmation of the Carolina slate belt as an exotic terrain: Science, v. 221, p. 649—651.
Snoke, A.W., Kish, S.A., Secor, D.T., Jr. (1980): Deformed Hercynian granitic rocks from the Piedmont of South Carolina: American Journal of Science, v. 280, p. 1019—1034.
Thomas, W.A. (1976): Evolution of Ouachita-Appalachian continental margin: Journal of Geology, v. 84, p. 323—342.
Thomas, W.A. (1983): Continental margins, orogenic belts, and intracratonic structures: Geology, v. 11, p. 270—272.
Tull, J.F. (1978): Structural development of the Alabama Piedmont northwest of the Brevard zone: American Journal of Science, v. 278, p. 442—460.
Williams, Harold (1978): Tectonic-lithofacies map of the Appalachians orogen: Memorial University, Newfoundland, scale 1:1,000,000.
Williams, Harold and Hatcher, R.D., Jr. (1983): Appalachian suspect terrains, in, Hatcher, R.D., Jr., Williams, Harold and Zietz, Isidore (eds.), Contributions to the Tectonics and Geophysics of Mountain Chains: Geological Society of America Memoir 158, p. 33—54.
Zietz, Isidore, Compiler (1982): Composite magnetic anomaly map of the United States Part A: Conterminous United States: Scale 1:2,500,000.

Manuscript received 31 Oct. 1983

Low-Grade Metamorphism in the Paratectonic Caledonides of the British Isles

R. E. Bevins
Department of Geology, National Museum of Wales, Cathays Park, Cardiff, Wales, CF1 3 NP, U.K.

G. J. H. Oliver/L. J. Thomas
Department of Geology, University of St. Andrews, Fife, Scotland, KY16 9ST, U.K.

Key Words

Low-grade metamorphism
Paratectonic Caledonides
Index Minerals
Conodont Colour Alteration
Illite Crystallinity

Abstract

The paratectonic zone of the Caledonide Orogenic belt in Britain and Ireland is commonly referred to as being non-metamorphosed. Recent work from many different parts of the paratectonic zone is presented here, which convincingly demonstrates that the area has suffered low-grade regional metamorphism ranging from the zeolite facies, through the prehnite-pumpellyite and pumpellyite-actinolite facies, into the greenschist facies with the localised development of blueschist and amphibolite facies. This metamorphic history is reflected in mineral growth in metavolcanic rocks and metagreywackes, conodont colour alteration patterns and the growth and degree of recrystallisation of clay minerals in pelitic rocks. The alteration histories of the different areas result from a variety of processes, including straightforward burial of sediments, tectonic burial in an accretionary prism, and hydrothermal and dynamothermal metamorphism associated with the generation and subsequent obduction of oceanic crust.

Introduction

Traditionally the Caledonian Orogenic belt of the British Isles has been divided into a northern metamorphic area and a southern non-metamorphic area (Read, 1961), corresponding to the orthotectonic and paratectonic zones of Dewey (1969) (see Fig. 1). The orthotectonic zone was subject to a complex phase of metamorphism and deforma-

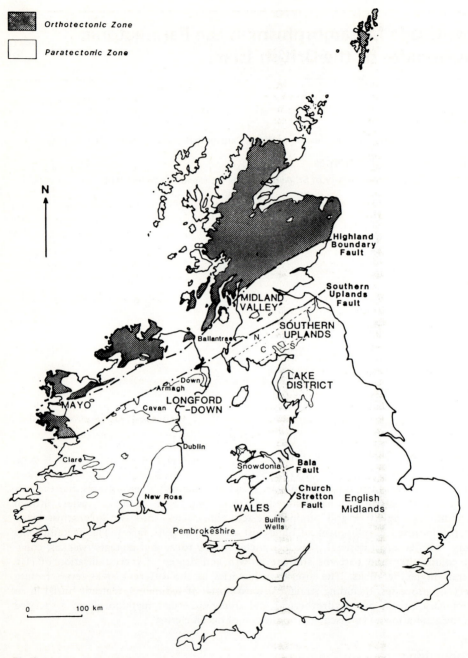

Fig. 1
Sketch map of the British Isles, showing the distribution of the paratectonic and orthotectonic zones.

tion during early Ordovician times (the Grampian Orogeny) and a range of metamorphic facies from low greenschist to upper amphibolite of a low to intermediate pressure facies series has been identified as a result of extensive studies, following the pioneer work of Barrow (1893, 1912). In contrast, systematic studies in the paratectonic zone have only recently commenced, as a result of the identification of critical secondary mineral assemblages in metavolcanic, metavolcaniclastic and metagreywacke rocks of this zone in County Mayo, Ireland (Ryan and Archer, 1977), Pembrokeshire, Wales (Bevins, 1978) and in County Cavan, Ireland (Oliver, 1978).

The contrasting metamorphic histories of the orthotectonic and paratectonic zones relate to the plate tectonic evolution of the Caledonian Orogenic belt during late Precambrian and Lower Palaeozoic times. Metamorphism in the orthotectonic zone probably resulted from compression of an ensialic basin sited between the foreland to the northwest and a continental ridge or landmass to the southeast at the onset of subduction associated with the closing of the Iapetus Ocean (Anderton, 1982). To the south, metamorphism in the paratectonic zone resulted largely from burial of sedimentary and volcanic sequences which had accumulated in a variety of environments associated with the evolution of Iapetus, including island arcs, fore-arc trenches and associated accretionary prisms, back-arc basins, in addition to the ocean floor itself. Specific examples of ocean floor hydrothermal and obduction-related metamorphism have also been identified. The constant reference to rocks of Cambrian through Silurian age of the paratectonic Caledonides as being non-metamorphosed (e.g. Dewey, 1969; Kelling, 1978; Anderson and Owen, 1980; Soper, 1980) has no doubt hindered the correct identification of metamorphic grade in this zone. The area is shown as being unmetamorphosed on the metamorphic map of Europe by Zwart and Sobolev, published in 1973, although the 1978 map depicts the area as belonging to the laumontite and prehnite-pumpellyite facies. Other estimates of metamorphic grade in this zone include that by Rast (1969), who thought that Wales had suffered zeolite facies metamorphism, and the much earlier suggestion by Woodland (1938) that the peculiar mineralogy of the manganiferous beds of the Harlech Dome region of Wales was due to chlorite-zone metamorphism. The earliest detailed investigations of rocks which have suffered this low grade of metamorphism were undertaken by Coombs (1954, 1960), working in South Island, New Zealand, whilst the pumpellyite-actinolite facies was defined by Hashimoto as recently as 1966. Despite this, however, there appears to have been a considerable number of descriptions of phases, particularly pumpellyite, characteristic of the prehnite-pumpellyite facies within the paratectonic zone of the British Caledonides (Bloxam, 1958; Nicholls, 1959; Kelling, 1962; Jenkins and Ball, 1964; Vallance, 1965), although prior to the present phase of work it was only Vallance (1965) and Coombs (1974) who attributed the presence of pumpellyite to low-grade metamorphism.

The identification of low-grade metamorphism is not solely dependant upon the recognition of critical mineral assemblages in metavolcanic rocks; other indicators include clay mineral assemblages and the degree of recrystallization of clay minerals in pelitic sediments (illite crystallinity), the temperature-induced colour change of conodonts (CAI value determinations), the reflectance and fixed carbon content of vitrinite particles in sediments and the calculation of % volatile matter in coals (VM values). There are, however, problems of correlation of 'grade' between the various techniques (see, for example, Kisch, 1974).

This paper presents data obtained over the past five years from rocks of Lower Palaeozoic age from various parts of the paratectonic zone of the Caledonide Orogen of the British Isles using a variety of techniques. Some of the data presented here is, as yet, unpublished, particularly that relating to the English Lake District. For detailed information concerning mineral data, the reader is referred to original publications.

The metamorphic history of the paratectonic Caledonides is described below in terms of different areas, which developed in contrasting tectonic environments during Lower Palaeozoic times. (See Figs. 1 to 6 for localities referred to in the text).

Wales

During much of Lower Palaeozoic time Wales was the site of a fault-bounded extensional basin, recently related by Kokelaar et al. (in press) to a marginal basin which developed behind an island arc system, located further to the north. Thick sequences of clastic sediments accumulated in the basin, associated with abundant contemporaneous volcanicity. On the adjacent shelf region, thinner, carbonate-dominated sediments accumulated in an area generally devoid of volcanic activity.

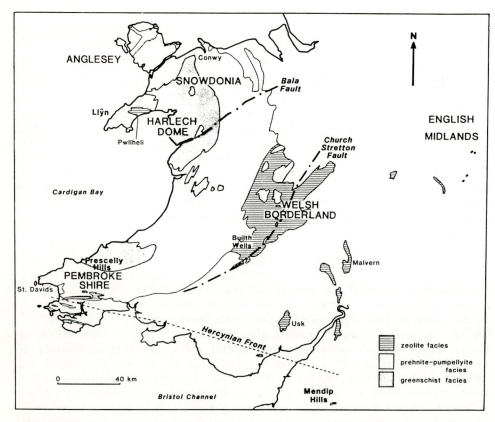

Fig. 2
Sketch map of Wales, the Welsh Borderland and the adjacent area of the English Midlands, showing the approximate distribution of the zeolite, prehnite-pumpellyite and greenschist facies. In view of their restricted extent, areas of pumpellyite-actinolite facies have not been included. The facies distributions are based on index minerals in metabasites, conodont colour alteration patterns and illite crystallinity; in view of this the boundaries are not intended to represent isograds.

The fist positive identification of low-grade metamorphism in Wales was made by Coombs (1974), who attributed the occurrence of pumpellyite in Ordovician volcanic rocks at Builth Wells to prehnite-pumpellyite facies metamorphism, in contrast to previous suggestions by Nicholls (1959) and Raam et al. (1969), who thought that it was of autometasomatic or deuteric origin, respectively. Following this, little attention appears to have been paid to what now proves to be the pervasive development of critical secondary mineral phases in the Welsh region until Bevins (1978) ascribed the presence of pumpellyite in Ordovician metabasites of N Pembrokeshire to prehnite-pumpellyite facies metamorphism. Roberts (1979) briefly reported the presence of prehnite, pumpellyite and stilpnomelane in Ordovician metabasic instrusive and extrusive rocks of the Conwy (northern Snowdonia) and Llŷn areas and interpreted the various assemblages as defining the prehnite-pumpellyite and pumpellyite-actinolite facies, passing locally into the greenschist facies. In more detail, Roberts (1981) described the distribution of three important isograds in Ordovician metabasic rocks of Snowdonia and Llŷn. He identified the pumpellyite-in, the pumpellyite-out/clinozoisite-in, and the biotite-in isograds, which allowed subdivision of the area into four metamorphic zones; the saponite and pumpellyite zones, which he considered to comprise the prehnite-pumpellyite facies in this area and the clinozoisite and biotite zones, belonging to the greenschist facies. The subsequent identification of analcime in veins at Gimlet Quarry, near Pwllheli (R.E. Bevins, unpublished data), in Roberts' saponite zone, however, suggests that this small area possibly belongs to the zeolite facies.

More recently, Bevins and Rowbotham (1983) have described in considerable detail the pervasive nature of secondary mineral assemblages within Ordovician metabasites of Wales and the Welsh Borderland. They reported extensive mineral data for prehnite, pumpellyite, actinolite, chlorite, albite, sphene, stilpnomelane, white mica and epidote, the distribution of which enabled the authors to suggest that most of the Ordovician sequences of Wales have suffered prehnite-pumpellyite facies metamorphism, although local patches of zeolite and greenschist facies grade occur (Fig. 2). The tendency for the areas of greenschist facies to be located in the more central parts of the basin, where thicker stratigraphical sequences occur, suggests that metamorphism resulted from burial of the rocks within the Welsh Basin. Unfortunately, the Ordovician sequences of the Welsh Borderland (east of the Church Stretton Fault) and of the English Midlands are largely devoid of volcanic rocks, which negates direct correlation of metamorphic facies of the basin with the shelf. Silurian volcanics in the Mendips Inlier, in Somerset, lying to the southeast of the basin, possess abundant laumontite and celadonite, with minor pumpellyite (R.E. Bevins, unpublished data), indicative of zeolite facies metamorphism in this area. However, this inlier lies to the south of the Hercynian Front, and the possibility that this assemblage results from Hercynian-age metamorphism cannot be ruled out.

Robinson et al. (1980) reported illite crystallinity data for rocks of Lower Cambrian to Westphalian age from Pembrokeshire. Those Lower Palaeozoic strata of the area affected solely by Caledonian metamorphism have Weaver values indicative of anchizone metamorphism, correlating well with the evidence from metabasites of that region, reported by Bevins (1978). Although three of the Cambrian samples collected from the St. David's area have crystallinities corresponding to the greenschist facies, Robinson et al. (op. cit.) did not suggest that this was the overall grade for Cambrian strata in this area. However, certain metabasic intrusions which invade Cambrian strata of this area possess well developed epidote-actinolite-chlorite-albite assemblages, suggesting that greenschist facies grade was reached at least locally.

Further illite crystallinity work has recently been undertaken, particularly in Central Wales and the Welsh Borderland. Bevins et al. (1981) reported high crystallinity values (indicative of the greenschist facies) in pelites of Cambrian and Ordovician age in the Harlech Dome region, to the north of the Bala Fault, and in Ordovician strata of the Prescelly Hills district (see Fig. 2). In contrast, the large tract of country to the south of the Bala Fault, in which strata of predominantly Upper Ordovician and Silurian age crop out, is dominated by anchizone crystallinities. The areas of greenschist facies crystallinity values correlate well with the areas possessing mineral assemblages in metabasites also diagnostic of the greenschist facies, described by Bevins and Rowbotham (1983). In the Welsh Borderland, in the shelf region, the preservation of kaolinite (D. Robinson and R.E. Bevins, unpublished data) is indicative of very low grades of metamorphism, with temperatures below 200 $^{\circ}$C (Dunoyer de Segonzac, 1970), probably no greater than 150 $^{\circ}$C.

The only published information to date concerning conodont colour alteration is that by Bergström (1980) for eleven samples from Ordovician strata in Wales and the Welsh Borderland. Samples collected from the basin area all have CAI values of 5 (i.e. 300 — 400 $^{\circ}$C), in sympathy with both metabasite mineral assemblage and illite crystallinity data. The four samples collected from the shelf area, to the east of the Church Stretton Fault, have markedly lower values, in the region 1 to 1.5 (i.e. ca. 100 $^{\circ}$C), which agrees with the unpublished clay mineral work (see above). Although based on only eleven samples, the same overall picture has recently been identified in a study utilising a much greater number of samples and sample sites (M.G. Bassett and N.M. Savage, pers. comm.). No published data exists for conodonts from the Silurian, but noticeably lower CAI values have recently been obtained from the area (R.J. Aldridge, pers. comm.) with low values, in the range 1.5 to 2.5, occurring in the shelf region and in the English Midlands (all to the east of the Church Stretton Fault), whilst those in the basin rise from 2.5 near the basin margin to about 4 in towards the centre of the basin.

An overall pattern of metamorphic grade is apparent (see Fig. 2). Cambrian to Silurian strata to the east of the Church Stretton Fault, in the shelf region, have suffered only very low grades of metamorphism. The lack of metabasites in this region precludes a facies description, but the occurrence of zeolites in Silurian metabasites in the Mendip Hills inlier and in Ordovician intrusions in the English Midlands (R.E. Bevins, unpublished data) strongly suggests that this area lies within the zeolite facies. To the west of the Church Stretton Fault a marked increase in grade is seen attributable to an increased thickness of overburden during burial, with Ordovician metabasites possessing mineral assemblages diagnostic of the prehnite-pumpellyite facies, passing locally, in the more central parts of the basin, into greenschist facies assemblages, a grade of metamorphism probably present in the Cambrian strata over much of the basin area. Metamorphic grade in the Silurian rocks of the basin passes from zeolite facies at the margin to prehnite-pumpellyite facies in towards the centre.

Assuming a geothermal gradient of 25 $^{\circ}$C km^{-1}, temperature estimates for the shelf region would imply Lower Palaeozoic stratigraphic thicknesses no greater than 6 km, whilst for the basin a maximum thickness of 15 km is implied for the areas of prehnite-pumpellyite facies, assuming that the boundary between the prehnite-pumpellyite facies and the greenschist facies occurs at approximately 375 $^{\circ}$C at 5 kb (see Schiffman and Liou, 1980). These estimates compare favourably with calculated maximum stratigraphic thicknesses of Lower Palaeozoic strata of 5 km for the shelf region and 13 km for the basin (see Kelling, 1978). However, it is very possible that the geothermal gradient in the Welsh Basin may have been in excess of 25 $^{\circ}$C km^{-1}, particularly in view of the marginal basin model outlined by Kokelaar et al. (in press). This requires further investigation.

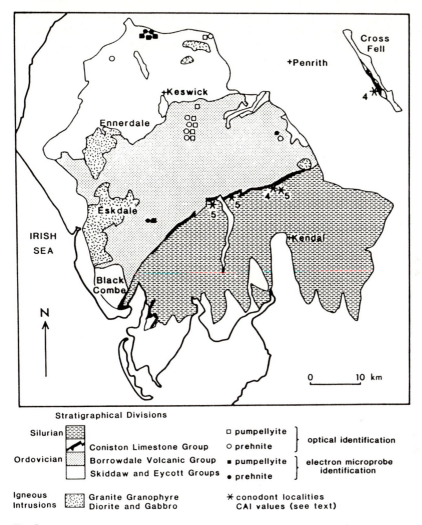

Fig. 3
Geological map of the Lake District showing the distribution of the major stratigraphic divisions, the distribution of pumpellyite and prehnite and conodont colour alteration index values (*) (the latter after Bergström, 1980 and R. J. Aldridge, pers. comm.).

The Lake District

The Lower Palaeozoic succession of the Lake District (see Fig. 3) is represented by an inlier of lower Ordovician (Arenig) greywackes and shales (the Skiddaw Group), conformably overlain to the north by tholeiitic transitional island arc to calc-alkaline lavas and associated sediments, belonging to the Eycott Group of Llanvirn age (Downie and Soper, 1972), in turn unconformably overlain to the south by the Borrowdale Volcanic Group. The age of the Borrowdale Volcanic Group, although not precisely known, lies between the upper Llanvirn and the middle Caradoc (Wadge, 1978). The group is dominated by andesitic lavas and dacitic to rhyolitic pyroclastics, chiefly ignimbrites, with an overall calc-alkaline chemistry, probably erupted in an island arc environment (Fitton and Hughes, 1970; Fitton et al., 1982). To the south, upper Ordovician and Silurian siltstones and sandstones overlap onto the older rocks. Granites emplaced during Ordovician and Silurian times intrude the Skiddaw Group, Borrowdale Volcanic Group and upper Ordovician and Silurian successions. A thorough review of Lake District geology is given by Moseley (1978), whilst a brief account of the volcanic sequences is given by Bevins et al. (this volume).

The extensive alteration of the volcanic rocks has been frequently referred to in descriptions of their petrology and as early as the account by Mitchell (1929) has been considered as a possible effect of regional metamorphism. Firman (1957) also referred to the possible regional metamorphic origin of epidote, whilst Oliver (1961) discussed the occurrence of the secondary minerals in terms of contact metamorphism in the region of the Eskdale Granite. Strens (1964) gave a detailed description of the relations, stability and conditions of formation of epidote and clinozoisite in the Borrowdale Volcanic Group. He divided the epidote occurrences into four types, viz. Shap-type veins, quartz-epidote veins, epidotic lavas and epidotic tuffs. He attributed the first two types to the effects of intrusions and the last two types to the effects of isochemical and alkali autometasomatism respectively. He suggested that autometasomatism occurred at temperatures between 300 °C and 500 °C. Helm (1970) described the Skiddaw Group of the Black Coombe Inlier as metasediments formed during low-grade metamorphism. Hitherto, there has been no report of minerals diagnostic of specific metamorphic facies in the Lower Palaeozoic rocks of the Lake District.

In the present study, compositions of the secondary phases in the volcanic rocks have been determined by electron microprobe analysis. It appears that secondary minerals have filled vesicles and formed from the partial or total breakdown and replacement of primary olivine, orthopyroxene, clinopyroxene, amphibole, Ca feldspar and glass to give assemblages of quartz ± sericite ± carbonate ± epidote ± clinozoisite ± albite ± sphene ± prehnite ± pumpellyite. Actinolite has been identified optically but from its restricted occurrence can nowhere be eliminated as possibly resulting from the effects of contact metamorphism by the associated granites.

The mineral phases in the secondary assemblages are apparently controlled by bulk-rock chemistry: epidote, clinozoisite, prehnite and pumpellyite are largely restricted to the more basic rocks (e.g. basaltic andesite) of the Eycott Group and the lower parts of the Borrowdale Volcanic Group whilst in the upper, more acidic parts of the Borrowdale Volcanic Group alteration products are confined to a non-diagnostic assemblage containing chlorite ± quartz ± carbonate ± sericite. Carbonate is a ubiquitous alteration phase and indicates a relatively high CO_2 fugacity (Zen, 1974) which possibly inhibited the development of index minerals. Relicts of primary feldspar compositions in phenocrysts suggest that metamorphic equilibrium was not attained.

Vein assemblages in the sedimentary successions are restricted to quartz ± chlorite ± carbonate although vein prehnite has been identified from the Embleton microdiorite which is itself intrusive into the Skiddaw Slate Group.

The coexistence of prehnite and pumpellyite and the absence of zeolites and actinolite is diagnostic of prehnite-pumpellyite facies metamorphism, as defined by Coombs (1960). Although autometasomatic alteration certainly occurred in the Lake District volcanic rocks, it is thought that the main alteration resulted from burial. Confirmatory evidence for this hypothesis is present in the Lake District sediments and their contained fossils (see below).

Thin section studies of the sedimentary sequences which occur above and below the two volcanic groups show an original detrital assemblage of quartz, feldspar and mica, associated with the secondary development of chlorite and white mica. Mafic minerals are scarce and the alteration has therefore produced no index minerals. X-ray diffraction study of the 2−6 μm fraction shows that the mica is illite, and metamorphic grade in these sediments can be defined by illite crystallinity, as expressed by the Hbrel index (Weber, 1972). The average Hbrel value for sediments of the Skiddaw Group is 144 (75 samples), with a range of 112−166, whilst that for upper Ordovician and Silurian sediments is 130 (40 samples) with a range of 99−150. These values are indicative of anchizone metamorphism and are compatible with the prehnite-pumpellyite facies identified in the Lake District volcanic successions, following the correlations of Weber (1972) and Kisch (1974).

CAI values of 5 (Bergström, 1980) and in the range 4−5 (R.J. Aldridge, pers. comm.) have been recorded from the upper Ordovician and Silurian sequences of the Lake District. According to Epstein et al. (1977) these values correspond to heating to between 300 °C and 400 °C (CAI 5) and 190 °C and 300 °C (CAI 4) which equate with temperatures of metamorphism within the prehnite-pumpellyite facies (Schiffman and Liou, 1980). The range of CAI values of the Silurian samples are indicative of slightly lower temperatures of metamorphism than the Ordovician samples, which is compatible with the model of burial metamorphism. Bergström (1980) reports a CAI value of 4 for the Ordovician of the Cross Fell Inlier (i.e. between 190 °C and 300 °C).

There is little certainty of regional metamorphic dates in the Lake District except for the growth of chlorite and illite along the cleavage which has been dated at 395 Ma (Ineson and Mitchell, in discussion in Wadge et al., 1974). Illite and chlorite growth in the Skiddaw Group and upper Ordovician and Silurian strata is found parallel to bedding as well as parallel to cleavage, which suggests growth during burial. The Skiddaw and Eycott groups have both suffered a complex history of burial and uplift. While metamorphism of the Skiddaw Group may have developed during burial by the Eycott Group, metamorphism of the Eycott Group must be a later event and probably resulted from overburden of the Borrowdale Volcanic Group. Metamorphism in the preserved units of the Borrowdale Volcanic Group may have been produced by burial from younger Borrowdale units, combined with the additional overburden of upper Ordovician and Silurian sedimentary units. There is no evidence to suggest that upper Ordovician and Silurian deposits overlay the Skiddaw and Eycott groups in the north, and metamorphism in these groups probably therefore occurred during Llanvirn to Caradoc times.

Southern Uplands and Longford-Down Massif

The Southern Uplands and Longford-Down Massif are underlain by greywackes and shales of Ordovician and Silurian age, which acccumulated in an accretionary prism developed on the northern side of the Iapetus Ocean (see below). Geographically, three distinct areas can be recognised, namely the Northern Belt, the Central Belt and the Southern Belt (see Figs 4 and 5).

The first published account of metamorphic rocks in the Southern Uplands was by the Geological Survey of Scotland in 1871 as part of the 'Explanation of Sheet 15'. The rocks of Bail Hill (Northern Belt) were described as a 'remarkable area of metamorphism'. Peach and Horne (1899) correctly re-interpreted these rocks as volcanic although recently McMurtry (in Hepworth et al., 1982) noted that the Bail Hill volcanics in fact do have a metamorphic history (see below).

Kelling (1962) reported prehnite veins in the pyroxene-bearing greywackes of the Portpatrick Group in the Rhinns of Galloway (Northern Belt). He also noted the occurrence of 'brown epidote' in the greywacke matrix but did not discuss the fact that pumpellyite

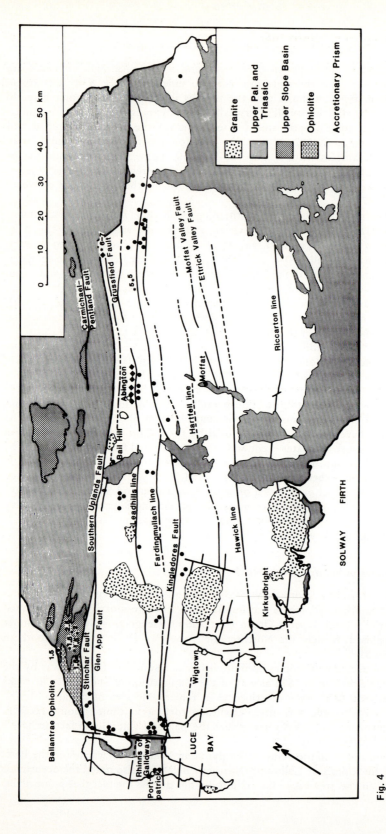

Fig. 4

Geological map of the Southern Uplands. Major reverse faults separate tracts with distinctive stratigraphic sequences which get younger from NW to SE (see Leggett et al., 1979 for details): beds within tracts young predominantly to the NW. Solid lines show major faults, dashed lines show inferred continuations from outcrops of basalt, chert and graptolitic shale (which occur in imbricate zones along the major faults): after sheets 1-11, 14-18, 24-26, 32-34 Geological Survey of Scotland. Southern (S), Central (C) and Northern (N) belts illustrated accordingly.

Sample localities: ◆ metabasalt and metadolerite with pumpellyite or prehnite and pumpellyite. ● basic-clast greywackes with prehnite and/or prehnite-pumpellyite veins. Conodont data from Bergström (1980); symbol as Figure 3.

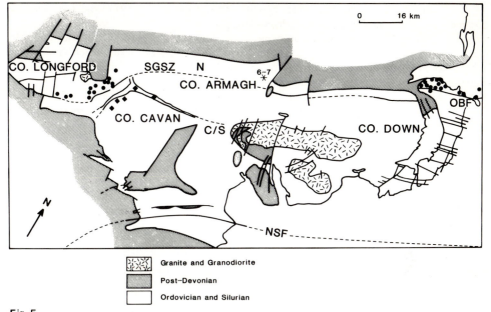

Fig. 5
Geological map of the Longford-Down Massif, after Leggett et al. (1979). Symbols same as Figure 4, except Co. Down localities, which are quartz + prehnite veins (L. Craig, pers. comm.).
SGSZ = Slieve Glah Shear Zone; OBF = Orlock Bridge Fault; NSF = Navan-Silvermines Fault.
N = Northern Belt, C/S = Central and Southern belts (undifferentiated).
Conodont data from Bergström (1980) with symbol as for Figure 3.

has very similar optical properties to epidote (see below). In 1978 Kelling still regarded the Southern Uplands as being unmetamorphosed.

Rust (1965) described the mineralogy of greywackes and slates in Wigtownshire (Central Belt). He identified illite, kaolinite and chlorite by X-ray diffraction and described the metamorphic growth of chlorite along the cleavage. However, he did not speculate on the grade of metamorphism.

Weir (1974) suggested that greywackes and slates of the Central Belt in Kirkudbrightshire had attained greenschist facies because X-ray diffraction studies revealed a lack of clay minerals but rather an assemblage of quartz, feldspar, calcite, chlorite and mica. However, Weir (1974) did not specify the kind of mica present and therefore the mineral parageneses could be attributed to zeolite through to greenschist facies. Watson (1976) studied the structural state of graphite and the reflectivity of graptolites in black shales near Moffat and on the east side of Luce Bay in the Central Belt. The results were used to suggest greenschist facies metamorphism; however, the reported values are also applicable to prehnite-pumpellyite facies metamorphism but not to the zeolite facies (Oliver and Leggett, 1980). None of these early researchers studied Southern Uplands rocks from a metamorphic point of view and it is to their credit that they reported these details at all.

In 1977 the Southern Uplands were interpreted as an accretionary prism formed at the active margin of the Iapetus Ocean (McKerrow et al., 1977; refined by Leggett et al.,

1979 to include the Longford-Down Massif). McKerrow et al. (1977) did not consider the metamorphic consequences of their model.

Oliver (1978) described rocks metamorphosed within the prehnite-pumpellyite facies in the Co. Cavan part of the Longford-Down Massif (Fig. 5). Weakly deformed sandstones, which contain abundant detrital quartz, feldspar and pyroxene and lesser amounts of detrital hornblende, biotite and white mica, are cut by veins containing prehnite, pumpellyite, calcite and quartz. Almost all feldspars are albitic. The abundance of veins with Ca-bearing minerals and the common occurrences of carbonate in the matrix of the sandstones were used by Oliver (1978) to suggest that albite had originated by *in situ* metamorphic breakdown of detrital Ca-plagioclase with the redistribution of calcium into the matrix and vein systems. Where the sandstones were not rich in detrital feldspar and clinopyroxene, the feldspars were albitised but there was no development of prehnite or pumpellyite veins, only carbonate veins.

Oliver (1978) described two sets of prehnite and pumpellyite veins; an early set of veins which are micro-folded about the cleavage and therefore predate the cleavage, and a second set which occur in shatter-veins which demonstrably cut across, and therefore post-date, the cleavage. The Drumcalpin Spilite Member (Oliver, 1978) and its extension to the east, the Carrickallen Member (O'Connor, 1975), are made up of tuff breccias containing spilite cobbles. The spilites show unaltered pyroxene phenocrysts in a matrix of fine-grained albite, sericite, celadonite, chlorite and sphene. Large phenocrysts of what were presumably calcic plagioclase are pseudomorphed by mixtures of albite, sericite and spongy pumpellyite. Prehnite occurs in rounded, radiating masses which either pseudomorph olivine crystals or infill former vesicles. Oliver (1978) suggested that straightforward burial caused the regional metamorphism although he did not preclude the effects of tectonic burial which might be found in an accretionary prism (McKerrow et al., 1977).

The concept of metamorphism in the Southern Uplands accretionary prism was discussed by Oliver and Leggett (1980). They noted the similarity of lithologies in Co. Cavan and the Southern Uplands and predicted that both regions had reached a similar metamorphic grade. Their investigations of the sandstone mineralogies confirmed the pattern found in Ireland. Sandstones rich in detrital feldspar and pyroxene consistently show albite replacement and abundant development of vein systems containing prehnite, pumpellyite, quartz and carbonate. Sample locations with the diagnostic vein assemblage are spread along the length of the Southern Uplands within the Northern and Central Belts. Basic-clast sandstones are not found in the Southern Belt, thus explaining the absence of prehnite and pumpellyite.

Oliver and Leggett (1980) investigated the Portpatrick Group in the Rhinns of Galloway (Kelling, 1962) and confirmed that both prehnite and pumpellyite occur with quartz and carbonate in veins. Electron microprobe analyses showed that the brown epidote-like material in the matrix of the basic-clast sandstones is in fact pumpellyite, although colourless epidote was also analysed in spongy inclusions in albite grains (the albite presumably being an alteration product of detrital calcic plagioclase). Colourless rounded detrital grains of epidote were also analysed in the matrix. Oliver and Leggett (1980) noted that most turbidite sandstones in the Southern Uplands show between 15% and 30% matrix composed of very fine-grained quartz, carbonate, chlorite and sericite and that the rocks can be classified as lithic greywackes. The abundance of matrix compared to recent deep sea turbidite sands (Hubert, 1964) and experimentally produced coarse-grained turbidites (Kuenen, 1966) suggests diagenetic and/or metamorphic recrystallisation of unstable detrital minerals. Oliver and Leggett (1980) noted the occurrence of rock clasts containing prehnite and pumpellyite and detrital grains and aggregates of prehnite in the Southern Uplands and suggested they could represent recycling of metamorphosed material, accreted earlier in the history of the accretionary prism (see below).

Bergström (1980) reported palaeotemperatures based on conodont colour in samples collected in the Southern Uplands and equivalent strata in Ireland. Two sample areas in the Northern Belt of Scotland and one locality from the Northern Belt of the Longford-Down Massif, in Co. Armagh, gave CAI values ranging from 5 to 7 suggesting temperatures higher than 300 °C (CAI 5) and possibly as high as 400 °C (CAI 7). These temperatures are compatible with experimentally-derived temperatures for the co-existence of prehnite and pumpellyite obtained by Schiffman and Liou (1980). Their data limit a 300 °C temperature to fluid pressures no less than 1.5 kb (ca. 5.5 km depth of burial). A reasonable geotherm would suggest 10 km depth of burial to produce 300 °C.

Hepworth et al. (1982) confirmed the petrographic results of Oliver and Leggett (1980) in the Abington area of the Southern Uplands and in addition reported illite crystallinity and vitrinite reflectance results. X-ray diffraction analyses of the <2 μm fraction of sandstones, siltstones and shales show quartz + feldspar + chlorite + illite mineralogies. Illite crystallinity values are of the anchizone of Kubler (1968) and comparable with illite crystallinity values found in prehnite-pumpellyite facies terrains elsewhere (Kisch, 1974). No systematic variation in illite crystallinity was found within or between fault blocks, suggesting that metamorphism continued after deformation. The maximum reflectance of graptolite fragments collected from the Abington area range from 3.91 to 8.22%, equivalent to the metaanthracite coal rank and the prehnite-pumpellyite facies (Kisch, 1974). These results show no aerial variation, agreeing with the illite crystallinity results.

McMurtry (in Hepworth et al., 1982) interpreted the middle Ordovician Bail Hill Volcanic Group as a mildly alkaline seamount which had formed in the pelagic environment of the Iapetus Ocean. Continued subduction of the Iapetus Ocean plate caused the volcano to be detached and accreted along with its enclosing trench sediments. The Bail Hill Volcanic Group contains original phenocrysts of calcic plagioclase and amygdales which contain mixtures of carbonate, thomsonite, prehnite, pumpellyite and albite. The matrix contains chlorite and pumpellyite, whilst veins of carbonate and analcime crosscut the volcanics. This mineral association demonstrates that the metamorphism occurred under zeolite facies conditions. McMurtry (ibid.) noted the lack of albitisation in volcaniclastics and the presence of kaolinite in siltstones in the same fault block as the zeolite-bearing volcanics. The stability maximum of analcime with quartz and kaolinite is about 200 °C (Dunoyer de Segonzac, 1970; Liou, 1971).

X-ray diffraction studies (G.J.H. Oliver et al., unpublished data) of <2 μm, 2–6 μm and <30 μm fractions of (dry, glycolated, heated, HCl and dimethylsulphoxide-treated) crushed sandstone and slate from Northern, Central and Southern Belts of Scotland and Ireland show a lack of kandite, smectite or vermiculite and the presence of chlorite, mica, quartz, calcite and feldspar. The mica peak positions and the half-height width of the (002) mica peak are appropriate for 2M illite. In Scotland, Hbrel values (Weber, 1972) of 317 samples collected on a regional scale lie between 90 and 160 with a strong mean at 110. In Co. Down, Co. Cavan and Co. Clare Hbrel values of 145 samples lie between 100 and 260 with a mean at 140, suggesting a slightly lower grade of metamorphism in Ireland. Both Ireland and Scotland have Hbrel values consistent with the prehnite-pumpellyite facies (Kisch, 1974).

Oliver and Leggett (1980) and Hepworth et al. (1982) used the accretionary prism concept of McKerrow et al. (1977) to model metamorphic processes in the Southern Uplands.

By association this model applies to the Longford-Down Massif as well. It was suggested that zeolite facies metamorphism of the Bail Hill Volcanic Group may have formed in three different ways; a) by hydrothermal-seawater alteration of lava immediately after extrusion; b) during burial in the trench environment, or c) by tectonic burial during accretion. There appears to be no criteria to separate the contribution from each of these processes. In any case, the block containing the volcano was never thrust down as deeply as the other blocks because the rest of the Southern Uplands — Longford-Down Massif appears to have been metamorphosed to prehnite-pumpellyite facies. Perhaps the positive topography of the volcano prevented deep accretion.

Following the discovery of prehnite-pumpellyite facies metamorphism in basalts dredged from the Vema Fracture Zone of the Mid-Atlantic Ridge (Mevel, 1981) it is conceivable that the Northern Belt spilites were metamorphosed to prehnite-pumpellyite facies prior to incorporation in the trench or accretionary prism. Certainly metamorphism continued during accretion since the associated sandstones are metamorphosed to prehnite-pumpellyite facies grade.

Burial metamorphism could have occurred in the trench where up to 3 km of sediment accumulated. However, the ca. 300 °C required for prehnite-pumpellyite facies metamorphism would not have been attained under the amount of cover and with the low heat flow found in trenches (Ernst, 1974). Tectonic burial in the accretionary prism under typical subduction zone geothermal gradients would produce 300 °C between 9 and 14 km of cover, depending on the gradient. If the trench was 5 km deep and the trench-slope emergent, then a minimum of an extra 4 km of cover of already accreted wedges would be required to initiate prehnite-pumpellyite facies metamorphism in the top of the wedge. The tectonic burial hypothesis is supported by the occurrence of prehnite-pumpellyite-bearing veins in sandstones which post-date the cleavage (which itself post-dates rotation of beds to the vertical) (Hepworth et al., 1982). Sedimentary burial metamorphism would have produced higher grades in the base of the ca. 3 km thick sedimentary pile in the trench. Therefore grade should increase southwards in each block of strata which has been rotated to the vertical. However, illite crystallinity and graptolite reflectance results show no such systematic variation within blocks and it is concluded that although metamorphism may have been initiated in the trench it continued with accretion and reached its peak after rotation. Thus metamorphism would have commenced in the Northern Belt in early Ordovician times, whilst in the Central and Southern Belts metamorphism would have commenced in Llandovery and Wenlock times respectively. Meanwhile rocks in the Northern Belt would have been rotated through the accretionary prism, reached their metamorphic peak and become emergent on the trench-slope break (Cockburnland of Walton, 1963) by the early Silurian, in time to provide detritus (including prehnite vein material) to the trench to the south. Thus metamorphism was diachronous in the Southern Uplands — Longford-Down Massif.

Midland Valley of Scotland

Published accounts of metamorphism in the Midland Valley are confined to the Highland Border Complex, exposed along the line of the Highland Boundary Fault (Fig. 1), and the Ballantrae Complex (Fig. 6). The Highland Border Complex is *either* Arenig age, post-Grampian, marginal basin oceanic crust with ocean islands, originally lying within the Midland Valley to the south of the Dalradian sedimentary basin (Longman et al., 1979; Curry et al., 1982); *or* the complex is (?)Cambrian age, pre-Grampian, marginal basin oceanic crust with ocean islands, originally lying within the Dalradian sedimentary basin

Low-Grade Metamorphism — British Isles

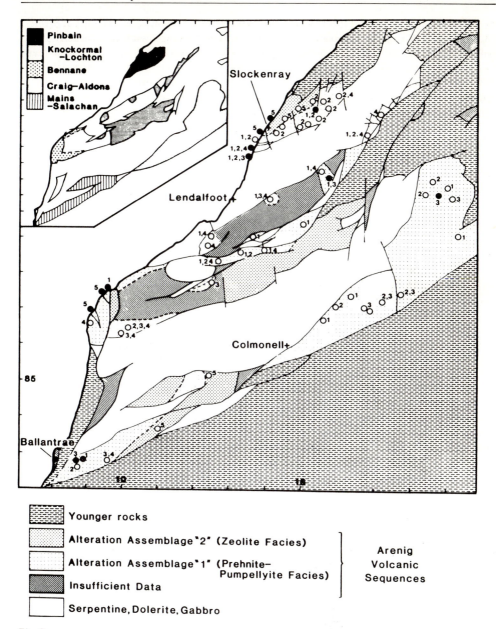

Fig. 6

Geological map of the Ballantrae Ophiolite Complex, showing the distribution of alteration mineral assemblages (see key).

○ optical identification, ● electron microprobe analysis identification.
1 = pumpellyite, 2 = prehnite, 3 = actinolite, 4 = epidote (all + albite, chlorite, sphene) and 5 = analcime, zeolite; + smectite ± albite, + chlorite + sphene.

Distribution of blocks referred to in the text appear in the inset. Blueschist and amphibolite localities are detailed by Bloxam and Allen (1960) and Treloar et al. (1980).

(Henderson and Robertson, 1982; Ikin, 1983). Whatever the origin, spilites found at Stonehaven, Glen Esk and Arran have a greenschist facies mineralogy, with albite + actinolite + chlorite + carbonate (Henderson and Fortey, 1982; Ikin, 1983). Hydrothermal metamorphism of the spilites may have occurred at the spreading centre of the marginal basin, although the greenschist facies grade of sediments associated with the spilites (albeit in faulted contact) suggests that regional metamorphism has affected the complex. According to Henderson and Robertson (1982) the timing of metamorphism was Grampian; Longman et al. (1979) and Curry et al. (1982) did not discuss the age but by inference it should be post-Grampian, perhaps, more precisely, pre-Caradocian (Longman et al., 1979). Evidence for obductive-emplacement of the Highland Border Complex is apparent in the garnet-hornblende schists seen on Bute (Ikin, 1983) and at Aberfoyle (Henderson and Fortey, 1982). These authors compare the garnet-hornblende schists with garnet amphibolites from the soles of tectonically-emplaced ophiolites e.g. Ballantrae (see below) and Newfoundland.

At Ballantrae several distinctive metamorphic environments are discernible. Blueschist facies rocks have been found, showing a transition to greenschist facies (Balsillie, 1937; Bloxam and Allen, 1960), which some workers (e.g. Coleman, 1972) would say implies a previous history of subduction. Balsillie (1937) considered the blueschists to be pre-Arenig in age. Jelínek et al. (1980) described igneous textures in dolerite dykes and gabbros of oceanic affinities which show metamorphic replacement by clinopyroxene and hornblende. They suggested that this high-grade metamorphism occurred whilst the rocks were still in the vicinity of the spreading ridge. Jelínek et al. (1980) also described chlorite, actinolite and biotite replacing the hornblende and pyroxene, and they assumed that this greenschist facies alteration was related to obduction. It is suggested here that this might equally have occurred when the ocean crust was still in the vicinity of the spreading ridge but at a distance where the heat flow was less. Serpentinisation and rodingitisation (Bloxam, 1954) may also have been initiated near the spreading ridge.

Obduction-related metamorphism has been recognised by Church and Gayer (1973), Spray and Williams (1980) and Treloar et al. (1980). Dynamothermal metamorphism produced a 50 m thick, downward succession of granulite, amphibolite and greenschist facies mylonite zones. These tectonites represent the aureole where a slab of hot ocean mantle was obducted onto lava-olistostrome. Serpentinisation (and ?rodingitisation) of the ocean slab continued after obduction, since serpentinite and not peridotite is now in contact with the aureole.

Spilitisation is ubiquitous in the Ballantrae lavas. Balsillie (1932, 1937) described albite, chlorite, prehnite, analcime, epidote, zoisite, quartz, calcite and pectolite as alteration minerals in spilites, gabbros and dolerite dykes. He considered the spilites resulted from metasomatic activity. Recently Bluck (1982) has reasoned convincingly that the metasomatism at Slockenray (Fig. 6) resulted from convective circulation of trapped sea water whilst the lava flows still retained their heat after crystallisation. Clasts of matched lava in the associated hyalotuffs are not spilitised; presumably because of their small size they lost heat very quickly.

Bloxam (1958) reported pumpellyite occurring in spilites in association with quartz, albite, calcite, chlorite and epidote. G.J.H. Oliver and J.L. Smellie (unpublished data) have shown that prehnite and pumpellyite (without zeolites or actinolite) occur in some spilites of the Pinbain and Bennane Head blocks, whilst other spilites have primary calcic plagioclase and analcime (alteration assemblages 1 and 2, Fig. 6, respectively). One sample of spilite has albite, chlorite, actinolite and pumpellyite but no zeolite or prehnite. Bonney's Dyke has, in addition to the prehnite and pectolite described by Balsillie

(1932), pumpellyite and actinolite replacing albite and clinopyroxene respectively (G.J.H. Oliver and J.L. Smellie, unpublished data). These analcime-prehnite-pumpellyite-actinolite assemblages suggest that burial metamorphism may have occurred at Ballantrae and that spilitisation may have been produced not just by hydrothermal sea water alteration (Coombs, 1974). It is possible that the obduction process piled up a sufficient sequence to initiate zeolite facies, prehnite-pumpellyite facies and pumpellyite-actinolite facies burial metamorphism in different parts of the complex. Bluck (1982) concluded that the northern spilites at Ballantrae are very likely oceanic island arc lavas and hyaloclastites. The total possible thickness of 7 km of these deposits should also contribute to burial metamorphism within the pile.

Radiometric dating of Ballantrae rocks indicates a relatively short Arenig history for ophiolite formation and obduction (Bluck et al., 1980). Trondhjemite and unfoliated gabbro in ophiolites are usually interpreted as having crystallised at a spreading ridge. Zircon ages indicate a crystallisation age for Ballantrae trondhjemite at 483 ± 4 Ma, whilst metamorphic amphibole from foliated gabbro has a K-Ar age of 487 ± 8 Ma (Bluck et al., 1980). The amphibole presumably dates cooling after the metamorphism associated with the spreading ridge (Jelínek et al., 1980) and is within the analytical uncertainties of the zircon age. Bluck et al. (1980) also dated amphibole from the dynamothermal aureole; the K-Ar 478 ± 4 Ma age is interpreted as a cooling age and indicates a very short time interval between magmatism, spreading centre metamorphism and obduction metamorphism (20 Ma allowing for decay constant and analytical uncertainties). Presumably, the burial metamorphism associated with the zeolite, prehnite-pumpellyite and pumpellyite-actinolite facies alteration was initiated shortly after obduction at 478 ± 4 Ma.

Bergström (1980) reported conodont CAI values of 1—2.5 for the Ordovician sediments unconformably overlying the Ballantrae Ophiolite Complex (Fig. 4). These values suggest burial temperatures of less than about 150 °C, and L.J. Evans (pers. comm.) has proven kaolinite in these sediments by X-ray diffraction analysis. R.J. Aldridge (pers. comm.) has identified conodonts from Silurian strata north of the Ballantrae Ophiolite Complex with CAI values of 2 (60—140 °C); these sediments also have kaolinite (R.M. Fowler, pers. comm.). Kaolinite is not stable above about 200 °C, (Dunoyer de Segonzac, 1970) and its presence therefore confirms the relatively low-temperature metamorphism of the cover to the Ballantrae Complex. R.J. Aldridge (pers. comm.) has identified conodonts with CAI values of 2.5, corresponding to temperatures of between 100 °C and 150 °C, in the Silurian of the Pentland Hills, an area where L.J. Evans (pers. comm.) has also found kaolinite in sediments of suitable composition.

It seems that the Lower Palaeozoic sediments of the Midland Valley have suffered a much lower grade of metamorphism than rocks of either the Southern Uplands accretionary prism or the Ballantrae Complex. This is compatible with the Midland Valley being the site of a back-arc basin in the early Ordovician, a volcanic and fore-arc basin in the middle and late Ordovician and an upper-slope basin in the Silurian.

Mayo, Western Ireland

The South Mayo Trough (Fig. 1) lies between the inferred extensions of the Highland Boundary Fault and the Southern Uplands Fault in W Ireland (Dewey, 1971; Phillips et al., 1976; Ryan and Archer, 1977; Leake et al., 1983). Therefore the trough could possibly be considered as belonging to the Midland Valley terrain. Leake et al. (1983) have recently presented evidence that the Dalradian rocks of Connemara were thrust southwards into their anomalous position in Llanvirn times although the precise nature of the contact with the South Mayo Trough is hidden by unconformable Silurian strata.

Ryan and Archer (1977) and Ryan et al. (1980) have made a preliminary study of metamorphism of basic volcanic rocks in the trough. They report palaeontological and geochemical data which suggests that the Lough Nafooey Group (the oldest in the trough) represents Tremadoc-age island arc volcanics. The volcanics are spilitised basalts and basaltic-andesites. The basal section has a greenschist facies mineralogy (quartz + albite + epidote + actinolite + zoisite), the bulk of the 2.6 km thickness has prehnite-pumpellyite facies assemblages (pumpellyite ± epidote ± prehnite without zeolites or actinolite), whilst laumontite occurs at the top of the volcanic succession indicating zeolite facies. The greenschist to prehnite-pumpellyite facies transition near the base of the volcanics indicates a temperature of $350-400\,°C$ (Schiffman and Liou, 1980). Bergström (1980) described conodonts with CAI values of 3.5 ($150-200\,°C$) from the Arenig Tourmakeady Limestone lying above the Lough Nafooey Group. This temperature is consistent with the stability of laumontite (Coombs et al., 1959). Natrolite and chabazite are zeolites reported in a dolerite sill and the neighbouring Llanvirn Partry Group sediments respectively (McManus, 1972). Oliver (1978) recognised pumpellyite in altered granophyre from the Curlew Mountains (Charlestown) an area which Ryan and Archer (1977) considered to be a northeasterly extension of the South Mayo Trough.

Ryan et al. (1980) used structural criteria to show that the Ordovician strata of the South Mayo Trough were subjected to burial metamorphism (more than 10 km of Ordovician flysch and molasse is present) and one fold phase prior to partial burial by Silurian strata. It seems likely that several metamorphic isograds could be mapped in the South Mayo Trough.

Southeastern Ireland

It has been assumed that the Lower Palaeozoic slates of SE Ireland have been regionally metamorphosed to greenschist facies. This is based on the regional occurrence of post-diagenetic chlorite and white mica (Wheatley, 1971; Shannon, 1978) and one occurrence of chloritoid + garnet at New Ross, in Co. Wexford (Fig. 1) (Shannon, 1978). However, slates with chlorite and unspecified white mica are not definitive of a particular facies (see discussion of the Southern Uplands, above). The significance of the supposed regional metamorphic chloritoid + garnet is equivocal. Brindley and Elsdon (1973) found apparently identical randomly orientated chloritoid porphyroblasts without garnet at Bunclody (50 km to the NE of New Ross along strike). They proposed that the porphyroblastic growth was a thermal effect of the Leinster Granite, exposed 2.5 km away. Shannon (1978) also found chloritoid (with andalusite) in schist at Oilgate which he related to a subsurface extension of the Ballynamuddagh Granite, exposed some 3 km distant. The New Ross locality is 8 km south of the southernmost outcrop of the Leinster Granite; it is not inconceivable that the granite extends southwards at a shallow depth just as at Oilgate. However, Shannon (1978) maintains that a comparison of the textural features at Bunclody and New Ross indicates that the New Ross chloritoid pre-dates the Leinster Granite.

G.J.H. Oliver and P.M. Shannon (unpublished data) have made a preliminary X-ray diffraction investigation of illite crystallinity of slates from SE Ireland. The samples were collected from a 10 km wide and 40 km long corridor running eastwards from New Ross to the east coast and cover the whole stratigraphic thickness of about 9 km (Shannon, 1978); chloritoid-bearing areas were avoided. The Hb_{rel} values show a regular increase from the Bray Group, at the base, to the top of the Ribband Group, consistent with a simple burial metamorphism mechanism (Fig. 7). The Duncannon Group samples, however, do not follow this pattern but have higher values. This group has an important

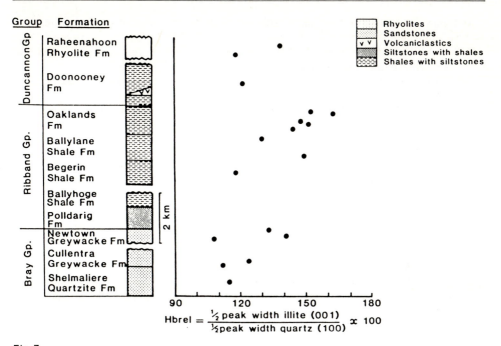

Fig. 7
Illite crystallinity results for SE Ireland. Stratigraphic column from Shannon (1978). Sample localities available from GJHO.

component of acid volcanics and related small scale intrusives, and it was not possible to collect samples which were possibly unaffected by heat from the lavas or intrusions. All the Hbrel values from SE Ireland slates indicate prehnite-pumpellyite facies (anchizone) grade metamorphism. Conodonts collected from the Tramore Limestone 30 km SSW of New Ross have CAI values of 5 (Bergström, 1980) which compares with values from prehnite-pumpellyite grade areas in the Southern Uplands and S Wales (see above). It appears, therefore, that the Lower Palaeozoic rocks of SE Ireland have suffered prehnite-pumpellyite facies metamorphism, which is supported by the identification of prehnite and pumpellyite in the Lambay Island Porphyry, exposed near Dublin (Oliver, 1978).

Conclusions and Summary

1) It is clear from this review of regional metamorphism in the paratectonic Caledonides of the British Isles that zeolite, prehnite-pumpellyite, pumpellyite-actinolite and greenschist facies occur on a regional scale. Blueschist and amphibolite facies rocks occur on a local scale.
2) The British paratectonic Caledonides can no longer be regarded as non-metamorphic. Indeed, the extent and facies distribution is now probably better defined in Britain than in the equivalent terrains in Newfoundland, NE Canada and USA. This is a reversal of the state of knowledge five years ago.

3) Metamorphic processes that have affected rocks of the British paratectonic Caledonides include:
   (a) hydrothermal metamorphism at ocean spreading ridges (e.g. the Ballantrae and Highland Border Complexes and the Southern Upland spilites),
   (b) serpentinisation and rodingitisation of oceanic mantle and lower crust (e.g. the Ballantrae and Highland Border Complexes),
   (c) hydrothermal autometamorphism of lavas erupted into sea water (e.g. the Ballantrae Complex and possibly other submarine lava occurrences),
   (d) dynamothermal metamorphism associated with obduction of ocean crust (e.g. the Ballantrae and Highland Border Complexes),
   (e) tectonic burial in an accretionary prism (e.g. the Southern Uplands and Longford-Down Massif),
   (f) sedimentary burial in volcanic arcs, inter-arc basins, fore-arc and back-arc basins (e.g. Wales, SE and W Ireland and the Midland Valley of Scotland).
4) This increase in knowledge of metamorphism has contributed significantly to the interpretation of the geological history of many areas in the British paratectonic Caledonides, particularly the Ballantrae and Highland Border Complexes.
5) Now that regional metamorphism of the British paratectonic Caledonides is firmly established, future work should concentrate on defining isograds in the various areas, with isotopic dating of suitable index minerals providing a detailed time frame for the different metamorphic episodes.
6) Finally, fluid inclusion and stable isotope studies might throw light on the nature of the fluids evolved and mostly lost as a consequence of the various paratectonic metamorphic events.

Acknowledgements

The authors would like to thank R.J. Aldridge, M.G. Bassett, L. Craig, P.N. Dunkley, L.J. Evans, J.G. Fitton, P.A. Floyd, R.M. Fowler, G.J. Lees, R.A. Roach, D. Robinson, G. Rowbotham, N.M. Savage, J.L. Smellie and R.J. Suthren for allowing us to utilise their unpublished results, providing thin sections and for fruitful discussions. G.J.H. Oliver and L.J. Thomas acknowledge receipt of NERC grants. Val Deisler and Jane Bidgood typed various versions of the text, whilst Lin Norton drafted the figures.

References

Anderson, J.G.C. and Owen, T.R. (1980): The structure of the British Isles (2nd edition). Pergamon Press, 162 pp.
Anderton, R. (1982): Dalradian deposition and the late Precambrian-Cambrian history of the N. Atlantic region: a review of the early evolution of the Iapetus Ocean. J. geol. Soc. Lond., 139, 421–434.
Balsillie, D. (1932): The Ballantrae Igneous Complex, south Ayrshire. Geol. Mag., 69, 107–131.
Balsillie, D. (1937): Further observations on the Ballantrae Igneous Complex, south Ayrshire. Geol. Mag., 74, 20–33.
Barrow, G. (1893): On an intrusion of muscovite-biotite gneiss in the southeast Highlands of Scotland and its accompanying metamorphism. Q. J. geol. Soc. Lond., 49, 330–358.
Barrow, G. (1912): On the geology of the lower Dee-side and the southern Highland border. Proc. Geol. Ass., 23, 268–284.
Bergström, S.M. (1980): Conodonts as palaeotemperature tools in Ordovician rocks of the Caledonides and adjacent areas in Scandinavia and the British Isles. Geol. Fören. Stockh. Förh., 102, 377–392.
Bevins, R.E. (1978): Pumpellyite-bearing basic igneous rocks from the Lower Ordovician of North Pembrokeshire. Min. Mag., 42, 81–83.

Bevins, R.E. and Rowbotham, G. (1983): Low-grade metamorphism within the Welsh sector of the paratectonic Caledonides. Geol. J., 18, 141–167.
Bevins, R.E., Robinson, D., Rowbotham, G., and Dunkley, P.N. (1981): Low grade metamorphism in the Welsh Caledonides. J. geol. Soc. Lond., 138, 634.
Bloxam, T.W. (1954): Rodingite from the Girvan-Ballantrae complex, Ayrshire. Min. Mag., 30, 525–528.
Bloxam, T.W. (1958): Pumpellyite from Ayrshire. Min. Mag., 31, 811–813.
Bloxam, T.W. and Allen, J.B. (1960): Glaucophane-schist, eclogite, and associated rocks from Knockormal in the Girvan-Ballantrae Complex, south Ayrshire. Trans. R. Soc. Edinburgh, 64, 1–27.
Bluck, B.J. (1982): Hyalotuff deltaic deposits in the Ballantrae ophiolite of SW Scotland: evidence for crustal position of the lava sequence. Trans. R. Soc. Edinburgh: Earth Sci., 72, 217–228.
Bluck, B.J. and Halliday, A.N. (1982): Age and origin of Ballantrae ophiolite and its significance to the Caledonian orogeny and the Ordovician time scale. [Reply to comment by Barrett et al., 1982]. Geology, 10, 331–333.
Bluck, B.J., Halliday, A.N., Aftalion, M., and MacIntyre, R.M. (1980): Age and origin of the Ballantrae ophiolite and its significance to the Caledonian orogeny and Ordovician time scale. Geology, 8, 494–495.
Brindley, J.C. and Elsdon, R. (1974): Chloritoid bearing rocks in the Leinster Massif. Sci. Proc. R. Dubl. Soc., 5A, 123–129.
Church, W.R. and Gayer, R.A. (1973): The Ballantrae ophiolite. Geol. Mag., 110, 497–510.
Coleman, R.G. (1972): Blueschist metamorphism and plate tectonics. 24th Intern. geol. Congr. Sect. 2, 19–26.
Coombs, D.S. (1954): The nature and alteration of some Triassic sediments from Southland, New Zealand. Trans. R. Soc. N.Z., 82, 65–109.
Coombs, D.S. (1960): Lower grade mineral facies in New Zealand. Rep. Int. geol. Congr. 21st Session, Norden, 1960, 13, 339–351.
Coombs, D.S. (1974): On the mineral facies of spilitic rocks and their genesis. In: Amstutz, G.C. (ed.): Spilites and spilitic rocks. Springer-Verlag, Berlin, 373–385.
Coombs, D.S., Ellis, A.J., Fyfe, W.S., and Taylor, A.M. (1959): The zeolite facies, with comments on the interpretation of hydrothermal syntheses. Geochim. Cosmochim. Acta, 17, 53–107.
Curry, G.B., Ingham, J.K., Bluck, B.J., and Williams, A. (1982): The significance of a reliable Ordovician age for some Highland Border rocks in central Scotland. J. geol. Soc. Lond., 139, 435–456.
Dewey, J.F. (1969): Evolution of the Appalachian/Caledonian orogen. Nature, 222, 124–129.
Dewey, J.F. (1971): A model for the Lower Palaeozoic evolution of the southern margin of the Early Caledonides of Scotland and Ireland. Scott. J. Geol., 7, 219–240.
Downie, C. and Soper, N.J. (1972): Age of the Eycott Volcanic Group and its unconformable relationship to the Skiddaw Slates in the English Lake District. Geol. Mag., 109, 259–268.
Dunoyer de Segonzac, G. (1970): The transformation of clay minerals during diagenesis and low-grade metamorphism: a review. Sedimentology, 15, 281–346.
Epstein, A.G., Epstein, J.B., and Harris, L.D. (1977): Conodont colour alteration – an index to organic metamorphism. Prof. Pap. U.S. geol. Surv., 995, 1–27.
Ernst, W.G. (1974): Metamorphism and ancient continental margins. In: Burk, C.A. and Drake, C.L. (eds): The Geology of Continental Margins. Springer-Verlag, New York, 933–950.
Explanation of Sheet 15 (1871): 'Geology of Dumfriesshire (North-west part), Lanarkshire (South part), Ayrshire (South-east part)'. Mem. geol. Surv. Scotland.
Firman, R.J. (1957): The Borrowdale Volcanic Series between Wastwater and the Duddon Valley, Cumberland. Proc. Yorks. geol. Soc., 31, 39–64.
Fitton, J.G. and Hughes, D.J. (1970): Volcanism and plate tectonics in the British Ordovician. Earth Planet. Sci. Lett., 8, 223–228.
Fitton, J.G., Thirlwall, M.F., and Hughes, D.J. (1982): Volcanism in the Caledonian orogenic belt of Britain. In: Thorpe, R.S. (ed.): Andesites: Orogenic andesites and related rocks. Wiley, 611–636.
Hashimoto, M. (1966): On the prehnite-pumpellyite-metagreywacke facies. J. geol. Soc. Japan, 72, 253–265.
Helm, D.G. (1970): Stratigraphy and structure in the Black Combe inlier, English Lake District. Proc. Yorks. geol. Soc., 38, 105–148.
Henderson, W.G. and Fortey, N.J. (1982): Highland Border rocks at Loch Lomond and Aberfoyle. Scott. J. Geol., 18, 227–245.
Henderson, W.G. and Robertson, A.H.F. (1982): The Highland Border rocks and their relation to marginal basin development in the Scottish Caledonides. J. geol. Soc. Lond., 139, 433–450.

Hepworth, B.C., Oliver, G.J.H., and McMurtry, M.J. (1982): Sedimentology, volcanism, structure and metamorphism of the northern margin of a Lower Palaeozoic accretionary complex — Bail Hill-Abington area of the Southern Uplands of Scotland. In: Leggett, J.K. (ed.): Trench-Forearc Geology. Spec. Publ. geol. Soc. Lond., 10, 521–533.

Hubert, J.F. (1964): Textural evidence for deposition of many north-western North Atlantic deep sea sands by ocean currents rather than turbidity currents. J. Geol., 72, 757–785.

Ikin, N.P. (1983): Petrochemistry and tectonic significance of the Highland Border Suite mafic rocks. J. geol. Soc. Lond., 140, 267–278.

Jelínek, E., Sovcek, J., Bluck, B.J., Bowes, D.R., and Treloar, P.J. (1980): Nature and significance of beerbachites in the Ballantrae ophiolite, SW Scotland. Trans. R. Soc. Edinburgh: Earth Sci., 71, 159–179.

Jenkins, D.A. and Ball, D.F. (1964): Pumpellyite in Snowdonian soils and rocks. Min. Mag., 33, 1093–1096.

Kelling, G. (1962): The petrology and sedimentation of Upper Ordovician rocks of the Rhinns of Galloway. Trans. R. Soc. Edinburgh, 65, 107–137.

Kelling, G. (1978): The paratectonic Caledonides. In: IGCP Project 27, Caledonian-Appalachian Orogen of the North Atlantic Region. Geol. Surv. Can. Pap. 78–13, 89–95.

Kisch, H.J. (1974): Anthracite and meta-anthracite coal ranks associated with 'anchimetamorphism' and 'very-low-stage' metamorphism, I, II, III. Proc. (K.) nederl. Akad. Wet., B, 77, 81–118.

Kokelaar, B.P., Howells, M.F., Bevins, R.E., Roach, R.A., and Dunkley, P.N. (in press): The Ordovician marginal basin of Wales. In: Kokelaar, B.P. and Howells, M.F. (eds): Marginal Basin Geology: Volcanic and associated sedimentary and tectonic processes in modern and ancient marginal basins. Spec. Publ. geol. Soc. Lond.

Kübler, B. (1968): La cristallinité de l'illite et les zones tout à fait supérieures du métamorphisme. Etages Tectoniques. A la Baconnière Neuchâtel, Switzerland, 105–122.

Kuenen, Ph.H. (1966): Matrix of turbidites; experimental approach. Sedimentology, 7, 267–297.

Leake, B.E., Tanner, P.W.G., Singh, D., and Halliday, A.N. (1983): Major southward thrusting of the Dalradian rocks of Connemara, western Ireland. Nature, 305, 210–213.

Leggett, J.K., McKerrow, W.S., Morris, J.H., Oliver, G.J.H., and Phillips, W.E.A. (1979): The northwestern margin of the Iapetus Ocean. In: Harris, A.L., Holland, C.H., and Leake, B.E. (eds): The Caledonides of the British Isles — reviewed. Spec. Publ. geol. Soc. Lond., 8, 499–512.

Liou, J.G. (1971): Analcime equilibria. Lithos, 4, 389–462.

Longman, C.D., Bluck, B.J., and Van Breemen, O. (1979): Ordovician conglomerates and the evolution of the Midland Valley. Nature, 280, 578–581.

McKerrow, W.S., Leggett, J.K., and Eales, M.H. (1977): Imbricate thrust model of the Southern Uplands of Scotland. Nature, 267, 237–239.

McManus, J. (1972): The stratigraphy and structure of the Lower Palaeozoic rocks of eastern Murrisk, Co. Mayo. Proc. Roy. Ir. Acad., 72B, 307–333.

Mevel, C. (1981): Occurrence of pumpellyite in hydrothermally altered basalts from the Vema fracture zone (Mid-Atlantic Ridge). Contr. Miner. Petrol., 76, 386–393.

Mitchell, G.H. (1929): The succession and structure of the Borrowdale Volcanic Series of Troutbeck, Kentmere and the western part of Longsleddale. Q. Jl geol. Soc. Lond., 85, 9–44.

Moseley, F. (1978): The Geology of the Lake District. Maney and Sons Limited, Leeds, 284 pp.

Nicholls, G.D. (1959): Autometasomatism in the Lower Spilites of the Builth Volcanic Series. Q. Jl geol. Soc. Lond., 114, 137–161.

O'Connor, E.A. (1975): Lower Palaeozoic rocks of the Shercock-Aghnamullen District, Counties Cavan and Monaghan. Proc. R. Ir. Acad., 75B, 499–530.

Oliver, G.J.H. (1978): Prehnite-pumpellyite facies metamorphism in County Cavan, Ireland. Nature, 274, 242–243.

Oliver, G.J.H. and Leggett, J.K. (1980): Metamorphism in an accretionary prism: prehnite-pumpellyite facies metamorphism of the Southern Uplands of Scotland. Trans. R. Soc. Edinburgh: Earth Sci., 71, 235–246.

Oliver, R.L. (1961): The Borrowdale Volcanic and associated rocks of the Scafell area, English Lake District. Q. Jl geol. Soc. Lond., 117, 377–417.

Peach, B.N. and Horne, J. (1899): The Silurian rocks of Britain, 1, Scotland. Mem. geol. Surv. Scotland, 749 pp.

Phillips, W.E.A., Stillman, C.J., and Murphy, T. (1976): A Caledonian plate tectonic model. J. geol. Soc. Lond., 132, 579–609.

Raam, A., O'Reilly, S.Y., and Vernon, R.H. (1969): Pumpellyite of deuteric origin. Am. Miner., 54, 320–324.

Rast, N. (1969): The relationship between Ordovician structure and volcanicity in Wales. In: Wood, A. (ed.): The Precambrian and Lower Palaeozoic rocks of Wales. Univ. Wales Press, Cardiff, 305–335.
Read, H.H. (1961): Aspects of Caledonian magmatism in Britain. Lpool Manchr geol. J., 2, 653–683.
Roberts, B. (1979): The geology of Snowdonia and Llŷn. Adam Hilger, Bristol, 183 pp.
Roberts, B. (1981): Low grade and very low grade regional metabasic Ordovician rocks of Llŷn and Snowdonia, Gwynedd, North Wales. Geol. Mag., 118, 189–200.
Robinson, D., Nicholls, R.A., and Thomas, L.J. (1980): Clay mineral evidence for low-grade Caledonian and Variscan metamorphism in south-western Dyfed, South Wales. Min. Mag., 43, 857–863.
Rust, B.R. (1965): The stratigraphy and structure of the Whithorn area of Wigtownshire, Scotland. Scott. J. Geol., 1, 101–133.
Ryan, P.D. and Archer, J.B. (1977): The South Mayo Trough: a possible Ordovician Gulf of California-type marginal basin in the west of Ireland. Can. J. Earth Sci., 14, 2453–2461.
Ryan, P.D., Floyd, P.A., and Archer, J.B. (1980): The stratigraphy and petrochemistry of the Lough Nafooey Group (Tremadocian), western Ireland. J. geol. Soc. Lond., 137, 443–458.
Schiffman, P. and Liou, J.G. (1980): Synthesis and stability relations of Mg-Al pumpellyite, $Ca_4Al_5MgSi_6O_{24}(OH)_7$. J. Petrol., 21, 441–474.
Shannon, P.M. (1978): The stratigraphy and sedimentology of the Lower Palaeozoic rocks of south-east County Wexford. Proc. R. Ir. Acad., 78B, 247–265.
Soper, N.J. (1980): Non-metamorphic Caledonides of Great Britain. Episodes, 1, 17–18.
Spray, J.G. and Williams, G.D. (1980): The sub-ophiolite metamorphic rocks of the Ballantrae Igneous Complex, SW Scotland. J. geol. Soc. Lond., 137, 359–368.
Strens, R.G.J. (1964): Epidotes of the Borrowdale Volcanic rocks of Central Borrowdale. Min. Mag., 33, 868–886.
Treloar, P.J., Bluck, B.J., Bowes, D.R., and Dudok, Arnost (1980): Hornblende-garnet metapyroxenite beneath serpentinite in the Ballantrae Complex of SW Scotland and its bearing on the depth provenance of obducted oceanic lithosphere. Trans. R. Soc. Edinburgh: Earth Sci., 71, 201–212.
Vallance, T.G. (1965): On the chemistry of pillow lavas and the origin of spilites. Min. Mag., 34, 471–481.
Wadge, A.J. (1978): Classification and stratigraphical relationships of the Lower Ordovician rocks. In: Moseley, F. (ed.): The Geology of the Lake District. Maney and Sons Limited, Leeds, 68–78.
Wadge, A.J., Harding, R.R., and Darbyshire, D.P.F. (1974): The Rubidium-Strontium age and field relations of the Threlked Microgranite. Proc. Yorks. geol. Soc., 40, 211–222.
Walton, E.K. (1963): Sedimentation and structure in the Southern Uplands. In: Johnson, M.R.W. and Stewart, F.H. (eds): The British Caledonides. Oliver and Boyd, Edinburgh, 91–97.
Watson, S.W. (1976): The sedimentary geochemistry of the Moffat Shales: a carbonaceous sequence in the Southern Uplands of Scotland. Unpublished Ph.D. Thesis, University of St. Andrews.
Weber, K. (1972): Kristallinität des Illits in Tonschiefern und andere Kriterien schwacher Metamorphose im nordöstlichen Rheinischen Schiefergebirge. N. J. Geol. Palaeont., 141, 333–363.
Weir, J.D. (1974): The sedimentology and diagenesis of the Silurian rocks on the coast west of the Gatehouse, Kirkudbrightshire. Scott. J. Geol., 10, 165–186.
Wheatley, C.J.V. (1971): Aspects of metallogenesis within the Southern Caledonides of Great Britain and Ireland. Trans. Instn Min. Metall., 80, B211–B224.
Woodland, A.W. (1938): Petrological studies in the Harlech Grit Series of Merionethshire. I. Metamorphic changes in the mudstones of the Manganese Shale Group. Geol. Mag., 75, 366–382.
Zen, E-an (1974): Prehnite- and pumpellyite-bearing mineral assemblages, west side of the Appalachian metamorphic belt, Pennsylvania to Newfoundland. J. Petrol., 15, 197–242.
Zwart, H.J. (1978): The metamorphic map of Europe. In: Zwart, H.J., Sobolev, V.S., and Niggli, E. (eds). Metamorphic Map of Europe: Explanatory text. Leiden.
Zwart, H.J. and Sobolev, V.S. (1973): Sub-Commission for the cartography of the Metamorphic Belts of the World. Leiden and UNESCO, Paris.

Revised manuscript received 27 Jan. 1984

A Review of Caledonian Volcanicity in the British Isles and Scandinavia

R. E. Bevins
Department of Geology, National Museum of Wales, Cardiff, Wales, U.K.

C. J. Stillman
Department of Geology, Trinity College, Dublin, Ireland

H. Furnes
Geologisk Institutt, Universitetet i Bergen, 5014 Bergen, Norway

Key Words

Caledonian
Volcanic rocks
British Isles
Scandinavia

Abstract

The Caledonian Orogenic Zone of the British Isles and Scandinavia contains abundant volcanic rocks which were largely generated as a consequence of the initial opening and subsequent closure of a major ocean, the Iapetus Ocean, in late Precambrian and early Palaeozoic times. In the British Isles volcanicity resulted principally from subduction of oceanic lithosphere northward beneath the Laurentian Plate and southward beneath the Southern Britain Microplate. In Scandinavia subduction, and sometimes obduction, of oceanic lithosphere beneath the Baltic Plate was responsible for the present day widespread distribution of Caledonian volcanic rocks. In contrast to the British Isles, however, the majority of the volcanic rocks in Scandinavia are allochthonous, occurring in major nappe units.

Introduction

Caledonian volcanicity is a product of the tectono-magmatic evolution of the Caledonian-Appalachian Orogen which, for the British Isles and Scandinavia, means the history of the Iapetus Ocean and its margins. On a global scale the orogen begins with the fragmentation, from about 650 Ma onwards, of a late Proterozoic super-continent (Piper, 1976), followed by the interaction of a number of continental or microcontinental plates, resulting in a series of small short-lived ocean basins, with associated subduction zones and attendant island arcs and marginal basins. The varying relative motions of these plates produced compressive or transpressive orogenic events, deforming and dislocating the sequences of rocks developed during Lower Palaeozoic times on the margins of these plates. The generation of significant volumes of magma, particularly during Ordovician times, is clearly related to these plate and microplate motions and recent studies of the volcanic sequences in the Caledonides of Britain, Ireland and Scandinavia have strongly influenced currently-accepted models relating to the evolution of the orogen.

Volcanicity appears to have occurred in a variety of environments, including island arcs and active continental margins, both with associated marginal basins, in addition to the true oceanic setting itself.

Caledonian volcanicity in the British Isles

The majority of the Caledonian volcanic rocks in Britain and Ireland were produced during closure of the Iapetus Ocean, resulting from subduction of oceanic lithosphere northward beneath a leading edge of the Laurentian Plate and southward beneath one or more microplates which underlie the southern parts of Britain and Ireland. The older sequences on the Laurentian Plate reveal to a certain extent the processes of rifting which generated the ocean basin and onto these were accreted the products of the supra-subduction magmatism which dominate the Caledonian sequences. Notably rare are obducted remnants of oceanic lithosphere itself which form such a significant part of the Scandinavian Caledonian volcanism (see below). Furthermore, such evidence of early activity is found only on the plate north of the suture that now marks the junction between the plates. This suture (Fig. 1), now largely concealed by Upper Palaeozoic and subsequent cover, can, with the aid of geophysical and structural analysis, be traced from the Shannon Estuary in SW Ireland northeastward via the Navan Fault to the Irish Sea, thence to the Solway Firth and across northern England (Phillips et al., 1976) although the recent work of Leake et al. (1983) questions the line of the suture in W Ireland. The different Proterozoic and Phanerozoic histories of the terrains on either side of the suture are mirrored in the volcanic rocks; hence it is appropriate to describe them separately. Localities referred to in the text are illustrated in Fig. 1.

1 Volcanicity South of the Suture

A deceptively simple geochemical zonation (Fitton and Hughes, 1970; Stillman and Williams, 1979) suggests that all the magmas in this region have been generated above a southward-dipping subduction system which may have changed its position in Ordovician times. Island arcs, marginal basins and active continental margins all developed from time to time in various places above this subduction system. However, this simplicity masks an early complexity, since the pre-Caledonian basements of SE Ireland, on the one hand, and Wales and S Britain on the other may well represent different terrains brought together and 'docked' at some unknown time. The latter basement is in fact rather young, representing a late Proterozoic accretion of island arc volcanics (Thorpe, 1979) with a dubious pre-Caledonian deformation history, whilst the former may be the most northerly representation of the Pentevrian Complex of northwest France and the Channel Islands (Phillips, 1981).

There is no evidence of early Iapetus Ocean crust but from the character of the earliest volcanic event, in Tremadoc times, it appears that Caledonian volcanism in this region was subduction-related from the start.

The volcanic rocks now seen nearest to the suture are found in northern England (Fig. 1). The Eycott Volcanic Group, of Arenig to Llanvirn age, crops out in the northern part

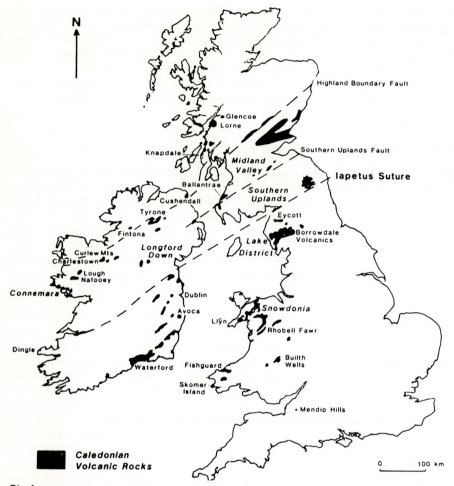

Fig. 1
Sketch map of the British Isles showing the distribution of Caledonian volcanic rocks, along with principal localities referred to in the text.

of the Lake District and rests conformably on the Skiddaw Slates. The volcanic rocks, some 2500 m thick, are largely submarine basalt and basaltic andesite lava flows (Millward et al., 1978). Relatively minor volumes of acidic lavas and tuffs occur, particularly towards the base of the group, indicating periodic explosive eruptions early in the development of the pile. The basic rocks show only a limited compositional range and are tholeiitic, considered by Fitton and Hughes (1970) to be transitional between island arc and calc-alkaline type, probably related to early-stage arc development. The Eycott lavas are similar to the basalts and basaltic andesites exposed in scattered inliers in the counties of Meath, Dublin and Kildare, in E Ireland (see Fig. 1), which were erupted somewhat later, building up sea mounts and oceanic islands with fringing reefs in an environment otherwise dominated by the accumulation of pelagic muds (Stillman and Williams, 1979).

Earliest eruptions in this area are of Llanvirn age, the most important events occurring in Caradoc and Ashgill times. A Llanvirn plinian pyroclastic eruption at Bellewstown represents the only acidic activity identified in this area.

To the south of the Eycott volcanics, occupying much of central and southern Lake District is the aerially extensive Borrowdale Volcanic Group of Llandeilo to Caradoc age. A total thickness of some 6000 m is dominated by andesitic lavas associated with dacitic to rhyolitic pyroclastic rocks, predominantly ignimbrites, with chemistry very similar to modern calc-alkaline suites (Fitton and Hughes, 1970; Fitton et al., 1982). The pile contains a significant proportion of reworked volcaniclastic debris, indicating an initial environment of deposition around a chain of stratovolcano islands composed of interbedded andesitic lavas and pyroclastics, with subsequent emergence coupled with continued activity producing a subaerial plateau (Millward et al., 1978).

In southeast Ireland during this time volcanicity was associated with the infilling of a major NE-SW trending basin which in Cambrian and earliest Ordovician times had seen the accumulation of sediments of the Bray and Ribband Groups (see Brück et al., 1979). Within this basin occur the 'coticule' beds, thought to be ocean floor metalliferous sediment related to volcanogenic-hydrothermal activity (P.S. Kennan, pers. comm.).

Evidence of the earliest volcanic episodes, of Llanvirn to Llandeilo age, is seen in coastal exposures in County Waterford, where submarine basalts of transitional tholeiitic character were erupted. These were followed by extrusion of calc-alkaline type basalts and basaltic andesites (Stillman et al., 1974; Stillman and Williams, 1979) which built up shield volcanoes and which were intimately associated with the emplacement of high-level sills into unconsolidated sediments. Following a short period of carbonate accumulation, volcanicity was renewed in Caradoc times with predominantly explosive acidic activity producing both subaerial and submarine ash-flows, along with the intrusion of rhyolite in the form of dykes, sills and domes. Some domes broke surface and were genetically related to the development of 'Kuroko-type' volcano-exhalative, copper-iron deposits, such as at Avoca (Sheppard, 1980). Rarer basaltic sheet lavas were erupted and this resulted in a characteristic bimodal association, with calc-alkaline subduction-related basic magmas and acidic rocks possibly derived from the partial melting of continental crust (Stillman and Williams, 1979).

It seems most likely therefore that the volcanic rocks of southeast Ireland and the Lake District were erupted in arcs developed on the southeastern margin of the Iapetus. However, a different tectonic setting exists for the area immediately to the south of this arc zone. Following a recent thorough re-examination of the Welsh Basin by Kokelaar et al. (in press) it is clear that the Welsh Basin developed by foundering of the underlying continental crust as a result of extensional tectonics, in a marginal basin environment. The basin was oriented NE-SW and active faults controlled both sedimentation and volcanism throughout the Lower Palaeozoic. During Cambrian times, marine sediments accumulated in the graben-like structure and evidence of volcanism is scant. The earliest important volcanic episode, exposed at Rhobell Fawr, occurred during late Tremadoc times. Kokelaar (1979) has identified the predominantly basaltic lavas as being of island arc-like character. The subduction system would then appear to have changed position in early Ordovician times, as the focus of arc-like magmatism moved northwards, to the southeast Ireland — Lake District zone and was replaced in Wales during Llanvirn to Caradoc times by predominantly tholeiitic volcanic activity, developed in a marginal basin environment (Bevins et al., in press). A distinct suite has been identified, with tholeiitic basalts occurring in association with rhyolitic lavas and pyroclastics, the latter most probably being related to partial melting of the underlying immature continental crust (Kokelaar et al., in press). Evidence for fractionation is sometimes present, as in

the Fishguard Volcanic Complex (Bevins, 1982), and it is possible that the rare intermediate lavas and a small proportion of the acidic lavas and pyroclastics result from differentiation of the voluminous basic magmas. The centres of volcanism migrated in both space and time, activity being of Arenig to Llanvirn age in SW Wales, predominantly pre-Caradoc in S Snowdonia, and of Caradoc age in N Snowdonia, whilst in eastern Wales (eg Builth Wells) and Llŷn evidence for both pre-Caradoc and Caradoc activity is present.
Intrusions are relatively common throughout Wales, generally being contemporaneous high-level sills, although rarer stock-like bosses occur. The granitoid intrusions of northern Llŷn have recently been described by Croudace (1982) who suggested that they were generated by crystal fractionation of andesitic liquids.
Throughout the region south of the suture the peak of volcanism was reached by the end of Ordovician times and from then on little Caledonian volcanism is recorded. Minor exceptions are the occurrences of rhyolitic and andesitic lavas and tuffs of Wenlock age in the far west of the Dingle Peninsula of SW Ireland (Parkin, 1976; Stillman, 1981), andesitic lavas in the Mendip Hills area of Somerset and the Tortworth area of Gloucestershire (Van de Kamp, 1969) and alkaline basalts on Skomer Island in SW Wales (Ziegler et al., 1969), all of Llandovery age, in addition to numerous bentonites of Silurian age in the Welsh Borderland (Ross et al., 1982) and Dyfed (Sanzen-Baker, 1972). The tectonic significance of these scattered outcrops is not however fully understood at present. The widespread and characteristic syn- and post-orogenic granite plutonism of the end-Caledonian orogenic phase appears to be the product of crustal melting together with a significant mantle-derived component (Halliday, 1984). However radiometric data now being obtained from the Leinster Granite suggest that melting may have commenced as early as late Ordovician to early Silurian times (P.S. Kennan, pers. comm.), which might suggest that by the end of the volcanic period the source magma had risen higher into the crust and that the subducted ocean slab and overlying mantle were no longer contributing magmas, a situation which was rather different from that on the northern side of the suture.

2 Volcanicity North of the Suture

No simple zonation is apparent here. The Laurentian Plate had a longer and more complex history involving interaction of both the Baltic Plate and the Southern Britain Microplate. This resulted in an early, Cambro-Ordovician (Grampian) and a later, end Silurian to mid Devonian (Caledonian) tectonic event, the latter being associated with extensive granite plutonism.
The earliest evidence of volcanicity on this northern plate is seen in the basaltic pillow lavas and associated intrusions of Dalradian (Cambrian) age, exposed in Knapdale, interpreted by Graham (1976) as tholeiitic magmas related to continental rifting, possibly coincident with the early development of the Iapetus Ocean. By Arenig times, small ocean basins fringed the southern margin of this plate. Bluck et al. (1980) have described an island arc-marginal basin association at Ballantrae which, together with oceanic crust, was obducted onto the continental margin between 470 and 490 Ma ago. However, a number of contrasting tectonic models have been forwarded to account for the various components of the Ballantrae Complex (see, for example, Church and Gayer, 1973; Wilkinson and Cann, 1974; Lewis and Bloxam, 1977; Barrett et al., 1982; Thirwall and Bluck, in press).
Recent palaeontological evidence (Curry et al., 1982) suggests that part of the Highland Border Complex, which contains debris derived from a back-arc basin to the south, is at

least of lower Arenig age, whilst along strike, in Ireland, the Tyrone Igneous Complex (Hartley, 1933) contains volcanic rocks and associated intrusions which may be related to a marginal basin or island arc environment. Possible equivalents of the Highland Border Complex occur even further to the W on Clare Island, County Mayo and comprise the Deer Park Complex (Ryan et al., 1983).

Evidence for active subduction has been described from Connemara where the Connemara migmatites contain an intrusive magmatic component which has been interpreted as the roots of an early Ordovician volcanic arc (Yardley et al., 1982). Slightly further to the north coeval submarine pillow lavas of Tremadoc age exposed at Lough Nafooey are island arc tholeiites, passing upwards into more evolved calc-alkaline lavas of Arenig age (Ryan et al., 1980). However it is unlikely that these little-deformed volcanics and pelagic sediments were produced in such close proximity to the Connemara orogenic zone and were probably only brought into contact with the Connemara rocks as late as Llanvirn times by some form of strike-slip movement or thrusting (see Leake et al., 1983). The volcanics developed on the southern margin of the South Mayo Trough which was a persistent feature during Ordovician and Silurian times. Bimodal acid and basic volcanism is similar to the Welsh Basin with the acidic magmas also probably related to crustal melting.

At Charlestown, 45 km to the northeast, similar volcanic rocks and marine sediments of Arenig age (Charlesworth, 1960) are associated with a porphyry-copper deposit which is possibly linked with a high-level, quartz-feldspar porphyry intrusion.

It appears that by early Ordovician times the southern margin of the Laurentian Plate was an active continental margin, with subduction-related arc magmatism and associated extensional basins.

To the south of this complex plate margin, the rocks presently exposed in the Southern Uplands of Scotland comprise a stacked, often inverted, sequence of sediments, volcaniclastics and lavas, interpreted by Leggett et al., (1979) as comprising ocean floor deposits of an accretionary wedge, obducted onto the continental margin. Support for this model is found in the lava geochemistry (Lambert et al., 1981) in addition to the presence of metalliferous sediments and cherts interpreted as having accumulated in an ocean floor environment.

Along strike, in the Longford-Down Massif, a similar sequence is present with greywackes, cherts and pelites of Llandeilo to Caradoc age associated with submarine basic lavas (Morris, 1981), which probably were generated in spreading marginal basins. Further southwest, in S Connemara, Ryan et al. (1983) have described further basaltic pillow lavas, also thought to be of Ordovician age.

There is some evidence for the development of an arc system to the north of the accretionary prism in later Ordovician and Silurian times. For example, derived clasts of arc-type rocks occur in the Ballantrae conglomerates (Bluck, 1978) and bentonites of possible arc-derivation are exposed in County Down (Cameron and Anderson, 1980). However, the complexity of the southern margin of the Laurentian Plate may well explain the lack of a clearly-defined chemical zonation as seen on the southern continental margin of the Iapetus Ocean.

By early Devonian times an epicontinental environment is seen, though the magmas continue to be calc-alkaline and to show a chemical zonation that may be related still to a northward-dipping subducted slab (Thirwall, 1981). In Ireland such volcanics are preserved only in a belt representing the westward extension of the Scottish Midland Valley. The Lower to Middle Old Red Sandstone volcanics of the Curlew Mountains are perhaps the equivalents of the Lower Old Red Sandstone volcanics in the northern part of the Midland Valley of Scotland. They are deformed to a higher degree than the overlying Lower Carboniferous sediments and may represent late-Caledonian activity. Andesitic lavas and tuffs show evidence of deposition in shallow water with drying-out features such as mud cracks and rain-drop pits seen on bedding surfaces in the tuffs. Nothing is yet known of their chemistry. Further northeast along strike towards the Midland Valley some 500 m of flow-banded andesites are seen above local basal ORS conglomerates in the Fintona block, 50 km NE of the Curlews. Finally at the Antrim coast in the Cushendall-Cushendun district andesite boulders and pyroclastic horizons are an important component of a clastic red-bed succession, possibly derived from a local vent now indicated by a quartz-andesite intrusive plug. There is also a series of feldspar-phyric dacite lavas. The structural position, correlated with the Scottish Midland Valley volcanics, and their petrography suggests that this magmatism may still be related to the supra-subduction activity on the continental plate north of the Iapetus Suture but more geochemistry is required to strengthen this suggestion.

In Scotland, volcanism associated with continental sediments of Old Red Sandstone facies, and broadly of Devonian age, occurs in three regions, namely the Lorne-Glencoe area (Groome and Hall, 1974; Roberts, 1974), the northern Midland Valley (Gandy, 1975; French et al., 1979), and the southern Midland Valley (Mykura, 1960). A wide range of rock types, from olivine basalts to rhyolitic ignimbrites, is present and the suite is dominantly calc-alkaline, similar to modern continental margin volcanics. Based on a regional variation in trace element concentrations and ratios, Thirlwall (1981) described the occurrence, during Lower and Middle Old Red Sandstone times, of a subduction zone which changed strike from ENE in Ireland and southern Scotland, to N in the North Sea area.

Caledonian volcanism is everywhere terminated by the main end-Caledonian orogenic event, during which plutonism replaced volcanism as the magmatic expression. Following the subsequent period of erosion the Hercynian cycle of activity produced an entirely different magmatism, associated with structures on the continental crust which now made up the British Isles.

Caledonian volcanicity in Scandinavia

Volcanic rocks of Caledonian age in Scandinavia occur within a number of nappe units, and evidence can be found of igneous activity associated with every stage of the Caledonian Orogen. As shown in Fig. 2 the various nappes have been divided into four major complexes, ie the Lower, Middle, Upper and Uppermost Allochthons (Roberts and Gee, in press). Most of the volcanic sequences are found in the Upper Allochthon, but some are also found in each of the others. The products of late Proterozoic crustal rifting may be found in the Lower, Middle and lower part of the Upper Allochthons; evidence of the spreading stages of the Iapetus Ocean are preserved in obducted ophiolite fragments found in the higher parts of the Upper and Uppermost Allochthons, and convergent plate margin products, indicating a complex sequence of arc, marginal basin and continental margin activity, are found in these same allochthons. The sequence indicates broadly that successive segments of the Iapetus Ocean crust and its Scandinavian supra-subduction margin have been brought eastward and stacked by a combination of obduction and tectonism in a succession of events which span the history of closing of the ocean.

The following account refers specifically to the volcanic rocks (including ophiolite fragments); to obtain a more complete picture of the overall magmatic evolution. the reader is referred to the comprehensive summary paper on Scandinavian Caledonian magmatism by Stephens et al. (in press).

Because of the nappe succession, in which stratigraphic packages are bounded by tectonic junctions, any meaningful systematic description of the volcanic rocks must be based on the tectonostratigraphy. Hence they will be dealt with in four sections as follows:

1 Rifting-Related Ensialic Igneous Activity in the Lower, Middle and Upper Allochthons

Within the Hedmark Group ("sparagmites") in the southeastern part of the Lower Allochthon (Fig. 2), minor occurrences of basalt lavas and dykes are found stratigraphically beneath the Moelv Tillite, and are considered to be of Upper Riphean or Vendian age (eg Saether and Nystuen, 1981). Geochemically these basalts are of typical continental tholeiitic composition, and are thought to have formed in an aulacogen related to the central rift zone of the Iapetus Ocean (Furnes et al., 1983b).

The Ottfjället Dolerites of the Särv Nappe (Middle Allochthon), cutting an unfossiliferous sequence of alluvial and shallow-marine feldspathic sandstones, dolomites and glaciogenic sediments (Røshoff, 1975; Kumpulainen, 1980), have been dated by the $^{40}Ar/^{39}Ar$ method to 665 ± 10 Ma (Claesson and Roddick, 1983). These are geochemically comparable to MORB and are thought to relate to the continental break-up and initial development stage of the Iapetus Ocean (Gee, 1975; Solyom et al., 1979a). In the upper part of the Leksdal Nappe (Middle Allochthon in the Tømmerås area, NE of Trondheim), equivalent to the Särv Nappe, metabasalt dykes with MORB/WPB affinities are thought to represent the same dyke generation as the Ottfjället Dolerites (Andreasson et al., 1979). The Kalak Nappe Complex (Sturt et al., 1978) of the Middle Allochthon in northern Norway, contains basic dykes in amphibolite facies with MORB to transitional continental WPB affinities (Gayer and Humphreys, 1981) which cut quartzo-feldspatic and pelitic schists.

Within the Seve Nappes, comprising the lower part of the Upper Allochthon (Zachrisson, 1973), amphibolites are associated with quartzo-feldspathic, garnet-mica and calcareous schists, and locally migmatitic gneisses (Trouw, 1973). The geochemical composition of these amphibolites indicates MORB affinity (Solyom et al., 1979b; Hill, 1980), and although their emplacement age has not been firmly established, Solyom et al. (1979b) favour a late Proterozoic age.

2 Arc-Related (Probably Ensimatic) Igneous Activity in the Køli Nappes and Equivalents of the Upper Allochthon

Lower Palaeozoic, fossil-bearing successions, occurring in a number of thrust nappes and generally metamorphosed in the greenschist facies, overlie the Seve Nappes and related units of the Upper Allochthon (eg Stephens et al., in press). In the central Caledonides these are referred to as the Køli Nappes, subdivided into the Lower, Middle and Upper Køli, and each nappe complex contains several volcanic-subvolcanic associations at different tectonostratigraphic and lithostratigraphic levels (Stephens, 1980). Only the upper part of the Upper Køli Nappe may be at the same tectonostratigraphic level as the Støren Nappe which contains ophiolite fragments. Hence, even though these arc-related volcanics certainly are younger than the oldest (group I) ophiolites, they logically will be described before the ophiolite, because of their lower tectonostratigraphic position.

In the Lower Køli Nappes volcanic rocks occur at two stratigraphic positions. The oldest, pre-Ashgill (probably mid Ordovician according to Strand and Kulling, 1972) volcanics are of a mixed basic and acid assemblage, associated with phyllites. The metabasalts are transitional tholeiitic/alkaline of WPB and MORB affinities (Stephens et al., in press).

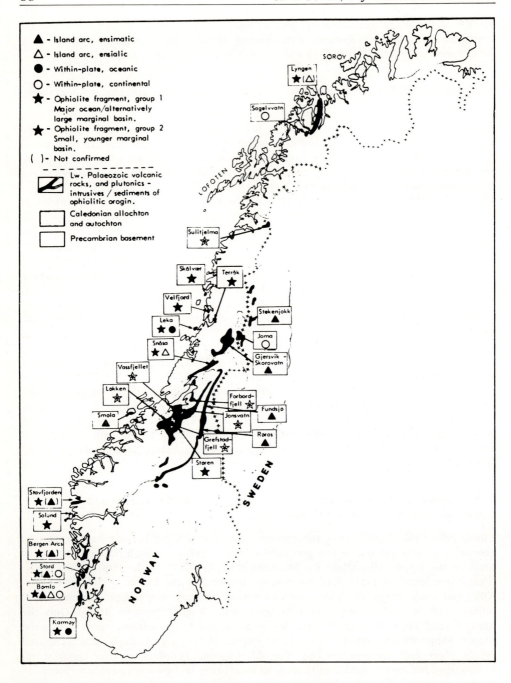

Fig. 2
Left-side map: The distribution of Lower Palaeozoic ophiolite fragments, island arc and within-plate volcanite associations in the Scandinavian Caledonides. Right-side map: Simplified (from Roberts and Gee, in press) tectono-stratigraphic map of the Scandinavian Caledonides.

Caledonian Volcanicity — British Isles and Scandinavia

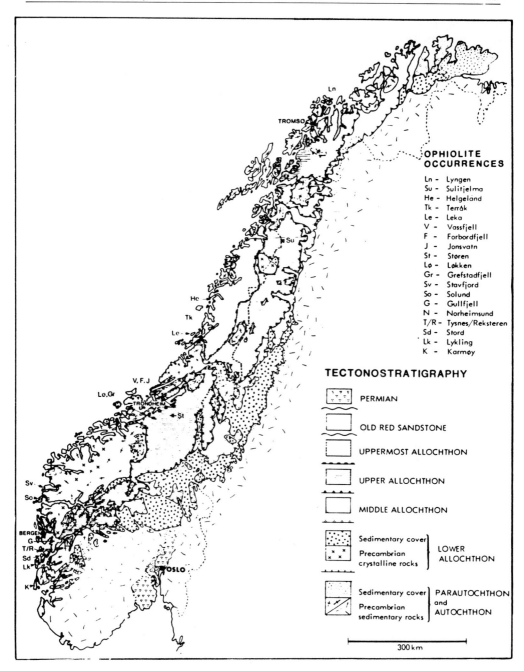

Fig. 2
For text refer to figure caption page 86

The youngest volcanics, interlayered with fossiliferous graphitic phyllites of Llandovery age (Stephens, 1977), are tholeiitic massive and pillow basalts of mainly MORB character.

The Middle Køli Nappes consist of several Ordovician (?) volcanic-subvolcanic complexes and calcareous flysch-like sediments (Zachrisson, 1964, 1969; Halls et al., 1977; Sjøstrand, 1978; Kollung, 1979; Lutro, 1979). The tectonostratigraphically lowest volcanics consist of tholeiitic to alkaline pillow lavas and hyaloclastites of MORB to WPB character (Stephens et al., in press), in association with phyllites and recrystallized ribbon cherts (Zachrisson, 1966; Olsen, 1980; Stephens et al., in press). The tectonostratigraphically highest volcanics within the Middle Køli, the Gjersvik Nappe, consist of a bimodal suite of basic and acid volcanics, of which the lowest basalts and basaltic andesites have trace element abundances similar to IAT, which are succeeded by tholeiitic MORB-like basalts, again overlain by alkaline basalts. This would point to a probably ensimatic island arc to a tensional environment related to arc splitting (Stephens, 1977, 1981, in press).

The Upper Køli Nappes contain basic volcanics at three tectonostratigraphic levels, of which the two lowest are of island arc tholeiitic composition, and have been interpreted as having formed in a fore-arc basin environment (Stephens and Senior, 1981). The uppermost basalts are of within-plate character (Stephens et al., in press).

3 Ophiolite Fragments and Associated Within-Plate Basalts in the Upper and Uppermost Allochthon

From Karmøy to Lyngen in the Norwegian Caledonides (Fig. 2), a number of rock sequences show geological features that suggest each of them as being part of an ophiolite assemblage. This is shown by the presence of a variably preserved, in some cases nearly complete, ophiolite pseudostratigraphy, including, in several cases, a conformable cap of pelagic sediments. The geochemical composition of the pillow lavas and dyke complexes, which predominantly show MORB characteristics, has also been emphasised. The ophiolite fragments occur in the upper part of the Upper Allochthon or in the Uppermost Allochthon (Leka, Skålvaer and Terråk fragments). Those recognized in the Støren Nappe, in the Trondheim region, are thought to lie either at the same tectonostratigraphic level (Roberts et al., 1970; Wolff and Roberts, 1980) or above (Gee and Zachrisson, 1974) the arc-related succession of the Køli Nappe Complex as described above.

A well-developed ultrabasic complex of layered, tectonized and variably serpentinized dunites, pyroxenites and periodotites, is represented on Leka (Fig. 2). The other ophiolite fragments contain only minor serpentinite bodies. Layered and high-level gabbros, sheeted dyke complexes and non-vesicular, sometimes variolitic pillow lavas are usually well developed. In the Støren and Stavfjorden ophiolites, as well as the Forbordfjell-Jonsvatn and Snåsa greenstones, however, only small gabbro bodies and occasional dykes are associated with the pillow lavas. Plagiogranites are usually present in the upper part of the gabbroic zone or the sheeted intrusive complex. Minor to substantial developments of pelagic or hemipelagic sediments, conformably overlying or intercalated with pillow lavas, occur in most of the sequences. Basic pillow lavas of alkaline, within-plate geochemical character, intercalated with pelagic sediments, are extensively developed on Karmøy, and possibly on Leka (Fig. 2). The detailed description of each individual ophiolite fragment and their geochemical development have been dealt with in a number of recent papers and have been summarized in Furnes et al. (1980, in press), Sturt et al. (in press) and Roberts et al. (in press, b).

The Norwegian ophiolites recognized to date range in age from probable Vendian to Middle Ordovician. An important bipartite grouping may be discerned from the combined criteria of field relationships, faunal constraints, radiometric dating and comprehensive geochemical studies. The two fundamental groups, I and II (Furnes et al., in press, b), equate in broad terms with Vendian (?) -Cambrian, and Ordovician ocean floor assemblages respectively, and the late Cambrian-earliest Ordovician Finnmarkian orogenic event (Sturt et al., 1978) provides a natural and convenient line of separation between the two.

Group I ophiolite complexes (Fig. 2) appear to be derivates of either a major ocean or a mature, arc-remote, marginal basin setting, and evidence points to their eastward obduction and internal deformation in Finnmarkian, pre-Middle Arenig times. There is also evidence for the existence of an early primitive island arc development of Upper Cambrian age (Furnes et al., in press, a). These island arc products were eventually obducted together with the subjacent oceanic crust during the Finnmarkian orogenic event. The Group I ophiolites and associated island arc developments were further dissected and deformed during the main Caledonian (Scandian), middle to late Silurian orogenic event.

Group II ophiolite complexes (Fig. 2) are restricted to Central Norway, and have been subjected to only the Scandian event. These ophiolite assemblages differ from those of Group I in being intra-Ordovician in age (late Arenig to Llanvirn and possibly Llandeilo), thus post-dating the Finnmarkian orogenesis. They also occur in association with arc-derived volcaniclastics as well as siliclastic sediments from two completely different sources, and generally show a more varied geochemical signature than the Group I ophiolites. The regional geological picture denotes that these Ordovician magmato-sedimentary associations, floored by a major unconformity and tectono-metamorphic break, accumulated in a fault-dissected, back-arc, marginal ocean basin milieu (Roberts et al., in press, a) with the Støren ophiolite slice and Gula Complex as part of the immediate substrate.

4 Arc-Related and Within-Plate Igneous Activity Associated with Ophiolites in the Upper Allochthon

The volcanites described here lie in the same tectonostratigraphic position as the ophiolite fragments, and they range in age from late Llanvirn to early Silurian. In the Trondheim region this island arc activity took place at about the same time as the development of the Group II ophiolites, and a genetic relationship is assumed.

A typical island arc development is provided by the late Arenig – early Llanvirn (Bruton and Bockelie, 1979) calc-alkaline volcanism (basalts to rhyolites) on the island of Smøla (Fig. 2) (Roberts, 1980). Roberts (1980) suggested that the Smøla island arc was built upon a basement of obducted Group I ophiolite. Time-equivalent island arc-related intrusives and extrusives in the Trondheim area are represented by the Hølonda andesites.

In southwest Norway a thick sequence of convergent plate-margin volcanics of Llandeilo to possibly early Silurian age is represented on the islands of Stord and Bømlo (Lippard, 1976; Nordås et al., in press). The earliest (Llandeilo) subaerial basalts to rhyolites of the Siggjo Complex, dated at around 460 Ma (Furnes et al., 1983a), are of mildly calc-alkaline to transitional tholeiitic/alkaline continental type ("Basin and Range" type). Upwards, in the post-Ashgill, the basalts are typically continental tholeiites. The stratigraphically highest unit on Bømlo (probably of Lower Silurian age), the Langevåg Group, shows pronounced changes in development from subaerial calc-alkaline basalts at the base, to submarine pillow basalts of alkaline and tholeiitic character towards the top of the sequence, indicating the incipient development of a marginal basin (Brekke et al., in press).

In N Norway, late Llandovery, shallow marine to subaerial basalts of mainly alkaline within-plate type are found in the Sagelvvatn area (Fig. 2).

Conclusions

In summary it seems that the Caledonian volcanism of Britain and Ireland can be related in simple terms to subduction systems operating during the closure of the Iapetus and the 'docking' of the Southern Britain Microcontinent (together with perhaps a fragment of NW France) with the southern flank of the Laurentian Plate southwest of its compressive collision with the Baltic Plate, and west of Tornquist's line (Fig. 3). Scandinavian Caledonian volcanism on the other hand was related to a quite different closure involving the Laurentian and Baltic plates. Whilst these two closures took place over the same general span of time, there is no direct correlation to be made between events taking place in the two systems. Indeed the history of closure and compressive collision of the Laurentian and Baltic plates, which resulted in the obduction of ocean crust, often developed in small marginal basins, followed by the stacking of the continental margin sequences in a series of nappes thrust one over the other, did not involve the crust which underlies southern Britain and Ireland. This apparently belongs to a microplate or plates which were detached from some other part of the African continent, which did not enter the 'Caledonian' orogenic belt until late in the process of closure (Anderton, 1982) and then docked largely by strike-slip motion. By this process collision and subsequent deformation was not always synchronous or compressive, the irregular outlines of the passing continents providing opportunities for the generation of pull-apart marginal basins and at a later stage the potential for high-level pluton emplacement where the thickened crust was wrench-faulted (see Hutton, 1982).

There is thus little basis for the direct correlation of volcanism in Scandinavia with that in Britain and Ireland, or even between that north and south of the Caledonian suture. Indeed it should be emphasised that the very nature of magmatic processes precludes

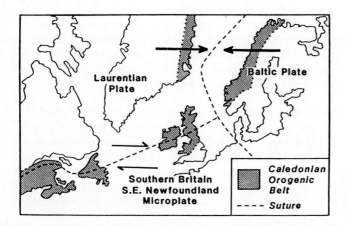

Fig. 3
Schematic map showing the probable distribution of plates and sutures following closure of the Iapetus Ocean.
(adapted from Anderton, 1982)

the use of volcanics, even though interbedded with sediments, as long-range chronostratigraphic correlatives. Volcanism is a process which can produce idential products wherever and whenever the environment is the same: similar tectono-magmatic settings can and do produce identical volcanic sequences in provinces which are geographically quite distinct and which formed at different times. An exception might be SW Britain where the opening and closing of Hercynian small ocean basins produced such rock sequences as the Lizard Complex. However these clearly belong to the Hercynian rather than the Caledonian cycle.

Acknowledgements

The authors would like to thank the many colleagues with whom they have, over the past few years, discussed aspects of Caledonian volcanicity. A review of this kind would not be possible without their invaluable contributions. Jane Bidgood is thanked for typing the text, whilst Ellen Irgens and Lin Norton provided cartographic assistance.

References

Anderton, R. (1982): Dalradian deposition and the late Precambrian-Cambrian history of the N. Atlantic region: a review of the early evolution of the Iapetus Ocean. J.geol.Soc.Lond., 139, 421–434.

Andreasson, P.G., Z. Solyom and D. Roberts (1979): Petrochemistry and tectonic significance of basic and alkaline-ultrabasic dykes in the Leksdal Nappe, Northern Region, Norway, Nor.geol. Unders., 348, 47–72.

Barrett, T.J., H.C. Jenkyns, J.K. Leggett and A.H.F. Robertson (1982): Comment and reply on 'Age and origin of Ballantrae ophiolite and its significance to the Caledonian orogeny and the Ordovician time scale'. Geology, 10, 331–333.

Bevins, R.E. (1982): Petrology and geochemistry of the Fishguard Volcanic Complex, Wales. Geol.J., 17, 1–21.

Bevins, R.E., B.P. Kokelaar and P.N. Dunkley (in press): Petrology and geochemistry of lower to middle Ordovician igneous rocks in Wales: a volcanic arc to marginal basin transition. Proc.Geol.Ass.

Bluck, B.J. (1978): Geology of a continental margin, 1: the Ballantrae Complex. In: D.R. Bowes and B.E. Leake (Editors): Crustal evolution in northwestern Britain and adjacent regions. Geol.J.Spec. Issue No. 10, 151–162.

Bluck, B.J., A.N. Halliday, M. Aftalion and R.M. MacIntyre (1980): Age and origin of Ballantrae ophiolite and its significance to the Caledonian orogeny and Ordovician time scale. Geology, 8, 494–495.

Brekke, H., H. Furnes, J. Nordås and J. Hertogen (in press): Lower Palaeozoic convergent plate margin volcanism on Bømlo, S.W. Norway, and its bearing on the tectonic environments of the Norwegian Caledonides. J. geol. Soc. Lon.

Brück, P.M., J.R.J. Colthurst, M. Feely, P.R.R. Gardiner, S.R. Penney, T.J. Reeves, P.M. Shannon, D.G. Smith and M. Vanguestaine (1979): Southeast Ireland: Lower Palaeozoic stratigraphy and depositional history. In: A.L. Harris, C.H. Holland and B.E. Leake (Editors): The Caledonides of the British Isles — reviewed. Spec.Publ.Geol.Soc.Lond., 8, 533–544.

Bruton, D. and J.F. Bockelie (1979): The Ordovician sedimentary sequence on Smøla, west central Norway. Nor.geol.Unders., 348, 21–31.

Cameron, T.D.J. and T.B. Anderson (1980): Silurian metabentonites in County Down, Northern Ireland. Geol.J., 15, 59–75.

Charlesworth, H.A.K. (1960): The Lower Palaeozoic inlier of the Curlew Mountains anticline. Proc.R. Ir.Acad., 61B, 37–50.

Church, W.R. and R.A. Gayer (1973): The Ballantrae Ophiolite. Geol.Mag., 110, 497–510.

Claesson, S. and J.C. Roddick (1983): $^{40}Ar/^{39}Ar$ data on the age and metamorphism of the Ottfjället Dolerites, Särv Nappe, Swedish Caledonides. Lithos, 16, 61–73.

Croudace, I.W. (1982): The geochemistry and petrogenesis of the Lower Palaeozoic granitoids of the Lleyn Peninsula, North Wales. Geochim.Cosmochim. Acta, 46, 609–622.

Curry, G. B., J. K. Ingham, B. J. Bluck and A. Williams (1982): The significance of a reliable Ordovician age for some Highland Border rocks in central Scotland. J.geol.Soc.Lond., 139, 435–456.
Fitton, J. G. and D. J. Hughes (1970): Volcanism and plate tectonics in the British Ordovician. Earth Planet.Sci.Lett., 8, 223–228.
Fitton, J. G., M. F. Thirlwall and D. J. Hughes (1982): Volcanism in the Caledonian orogenic belt of Britain. In: R. S. Thorpe (Editor): Andesites: Orogenic andesites and related rocks. Wiley, 611–636.
French, W. J., M. D. Hassan and J. Westcott (1979): The petrogenesis of Old Red Sandstone volcanic rocks of the western Ochils, Stirlingshire. In: A. L. Harris, C. H. Holland and B. E. Leake (Editors): The Caledonides of the British Isles – reviewed. Spec. Publ. Geol. Soc. Lond., 8, 635–642.
Furnes, H., H. Austrheim, K. G. Amaliksen and J. Nordås 1983a: Evidence for an incipient early Caledonian (Cambrian) orogenic phase in southwest Norway. Geol. Mag., 120, 607–612.
Furnes, H., J. P. Nystuen, A. O. Brunfelt and S. Solheim (1983b): Geochemistry of Upper Riphean – Vendian basalts associated with the "sparagmites" of southern Norway. Geol. Mag., 120, 349–361.
Furnes, H., D. Roberts, B. A. Sturt, A. Thon and G. H. Gale (1980): Ophiolite fragments in the Scandinavian Caledonides. Proc. Int. Ophiolite Symp. Cyprus 1979, 582–600.
Furnes, H., P. D. Ryan, T. Grenne, D. Roberts, B. A. Sturt and T. Prestvik (in press): Geological and geochemical classification of the ophiolite fragments in the Scandinavian Caldeonides. In: D. G. Gee and B. A. Sturt (Editors): The Caledonide Orogen – Scandinavia and Related Areas. New York, John Wiley.
Gandy, M. K. (1975): The petrology of the Lower Old Red Sandstone lavas of the eastern Sidlaw Hills, Perthshire, Scotland. J.Petrology, 16, 189–211.
Gayer, R. A. and R. J. Humpreys (1981): Tectonic modelling of the Finnmark and Troms Caledonides based on high level igneous rock geochemistry. Terra cognita, 1, 44.
Gee, D. G. (1975): A tectonic model for the central part of the Scandinavian Caledonides. Am.J.Sci., 275-A, 468–515.
Gee, D. G. and E. Zachrisson (1974): Comments on stratigraphy, faunal provinces and structure of the metamorphic allochthon, central Scandinavian Caledonides. Geol.Fören.Stockh.Förh., 96, 61–66.
Graham, C. M. (1976): Petrochemistry and tectonic significance of Dalradian meta-basaltic rocks of the SW Scottish Highlands. J.geol.Soc.Lond., 132, 61–84.
Groome, D. R. and A. Hall (1974): The geochemistry of the Devonian lavas of the northern Lorne plateau, Scotland. Min.Mag., 39, 621–640.
Halliday, A. N. (1984): Coupled Sm-Nd and U-Pb systematics in late Caledonian granites and the basement under northern Britain. Nature, 307, 229–233.
Halls, C., A. Reinsbakken, I. Ferriday, A. Haugen and A. Rankin (1977): Geological setting of the Skorovas orebody within the allochthonous volcanic stratigraphy of the Gjersvik Nappe, central Norway. In: I. Gass (Editor): Volcanic Processes in Ore Genesis. Spec.Publ.Geol.Soc.Lond., 7, 128–151.
Hartley, J. J. (1933): The geology of north-eastern Tyrone and the adjacent portions of County Londonderry. Proc.R.Ir.Acad., 41B, 218–285.
Hill, T. (1980): Geochemistry of the greenschists in relation to the Cu-Fe deposits in the Ramundberget area, central Swedish Caledonides. Nor.geol.Unders., 360, 195–210.
Hutton, D. H. W. (1982): A tectonic model for the emplacement of the Main Donegal Granite, NW Ireland. J.geol.Soc.Lond., 139, 615–632.
Kollung, S. (1979): Stratigraphy and major structures of the Grong District, Nord-Trøndelag. Nor. geol.Unders., 354, 1–51.
Kokelaar, B. P. (1979): Tremadoc to Llanvirn volcanism on the southeast side of the Harlech dome (Rhobell Fawr), N. Wales. In: A. L. Harris, C. H. Holland and B. E. Leake (Editors): The Caledonides of the British Isles – reviewed. Spec.Publ.Geol.Soc.Lond., 8, 591–596.
Kokelaar, B. P., M. F. Howells, R. E. Bevins, R. A. Roach and P. N. Dunkley (in press): The Ordovician marginal basin of Wales. In: B. P. Kokelaar and M. F. Howells (Editors): Marginal Basin Geology: Volcanic and associated sedimentary and tectonic processes in modern and ancient marginal basins. Spec.Publ.Geol.Soc.Lond.
Kumpulainen, R. (1980): Upper Proterozoic stratigraphy and depositional environments of the Tossåsfjället Group, Särv Nappe, southern Swedish Caledonides. Geol.Fören.Stockh.Förh., 102, 531–550.
Lambert, R. St. J., J. G. Holland and J. K. Leggett (1981): Petrology and tectonic setting of some Ordovician volcanic rocks from the Southern Uplands of Scotland. J.geol.Soc.Lond., 138, 421–436.
Leake, B. E., P. W. G. Tanner, D. Singh and A. N. Halliday (1983): Major southward thrusting of the Dalradian rocks of Connemara, western Ireland. Nature, 305, 210–213.

Leggett, J. K., W. S. McKerrow and M. N. Eales (1979): The Southern Uplands of Scotland, a Lower Palaeozoic accretionary prism. J.geol.Soc.Lond., 136, 755—770.
Lewis, A. D. and T. W. Bloxam (1977): Petrotectonic environment of the Girvan-Ballantrae lavas from rare-earth element distributions. Scott.J.Geol., 13, 211—222.
Lippard, S. L. (1976): Preliminary investigations of some Ordovician volcanics from Stord, West Norway. Nor.geol.Unders., 327, 41—66.
Lutro, O. (1979): The geology of the Gjersvik area, Nord-Trøndelag. Nor.geol.Unders., 354, 53—100.
Millward, D., F. Moseley and N. J. Soper (1978): The Eycott and Borrowdale volcanic rocks. In: F. Moseley (Editor): The geology of the Lake District. Maney and Sons Limited, Leeds, 99—120.
Morris, J. H. (1981): The geology of the western end of the Lower Palaeozoic Longford-Down inlier, Ireland. Unpublished Ph.D. Thesis, University of Dublin.
Mykura, W. (1960): The Lower Old Red Sandstone igneous rocks of the Pentland Hills. Bull.geol.Surv. Gt. Br. 16, 131—155.
Nordås, J., K. G. Amaliksen, H. Brekke, R. Suthren, H. Furnes, B. A. Sturt and B. Robins (in press): Lithostratigraphy and petro-chemistry of Caledonian rocks on Bømlo, S.W. Norway. In: D. G. Gee and B. A. Sturt (Editors): The Caledonide Orogen — Scandinavia and Related Areas. New York, John Wiley.
Olsen, J. (1980): Genesis of the Joma stratiform sulphide deposit, central Norwegian Caledonides. Proc. 5th IAGOD symposium, Alta, Utah 1978, 1, 745—757.
Parkin, J. (1976): Silurian rocks of Inishvikillane, Blasket Islands, County Kerry. Sci.Proc.R.Dubl. Soc., 5A, 277—291.
Phillips, W. E. A. (1981): The Pre-Caledonian Basement. In: C. H. Holland (Editor): A geology of Ireland. Scottish Academic Press, Edinburgh, 7—16.
Phillips, W. E. A., C. J. Stillman and T. Murphy (1976): A Caledonian plate tectonic model. J.geol.Soc. Lond., 132, 579—609.
Piper, J. D. A. (1976): Palaeomagnetic evidence for a Proterozoic supercontinent. Phil.Trans.R.Soc. Lond., 280A, 469—490.
Roberts, D. (1980): Petrochemistry and palaeogeographic setting of Ordovician volcanic rocks of Smøla, Central Norway. Nor.geol.Unders., 359, 43—60.
Roberts, D., J. Springer and F. Chr. Wolff (1970): Evolution of the Caledonides in the northern Trondheim region, Central Norway: a review. Geol.Mag., 107, 133—145.
Roberts, D. and D. G. Gee (in press): Caledonian tectonics in Scandinavia. In: D. G. Gee and B. A. Sturt (Editors): The Caledonide Orogen-Scandinavia and Related Areas. New York, John Wiley.
Roberts, D., T. Grenne and P. D. Ryan (in press, a): Ordovician marginal basin development in the central Norwegian Caledonides. In: B. P. Kokelaar and M. F. Howells (Editors): Marginal Basin Geology: Volcanic and associated sedimentary and tectonic processes in modern and ancient marginal basins. Spec.Publ.Geol.Soc.Lond.
Roberts, D., B. A. Sturt and H. Furnes (in press, b): Volcanic assemblages and environments in the Scandinavian Caledonides and the sequential development history of the mountain belt. In: D. G. Gee and B. A. Sturt (Editors): The Caledonide Orogen-Scandinavia and Related Areas. New York, John Wiley.
Roberts, J. L. (1974): The evolution of the Glencoe Cauldron. Scott.J.Geol., 10, 269—282.
Røshoff, K. (1975): A possible glaciogene sediment in the Särv Nappe, central Swedish Caledonides. Geol.Fören.Stockh.Förh., 97, 192—195.
Ross, R. J., Jr., C. W. Naeser, G. A. Izett, J. D. Obradovich, M. G. Bassett, C. P. Hughes, L. R. M. Cocks, W. T. Dean, J. K. Ingham, C. J. Jenkins, R. B. Rickards, P. R. Sheldon, P. Toghill, H. B. Whittington and J. Zalasiewicz (1982): Fission-track dating of British Ordovician and Silurian stratotypes. Geol.Mag., 119, 135—153.
Ryan, P. D., P. A. Floyd and J. B. Archer (1980): The stratigraphy and petrochemistry of the Lough Nafooey Group (Tremadocian), western Ireland. J.geol.Soc.Lond., 137, 443—458.
Ryan, P. D., V. K. Sawal and A. S. Rowlands (1983): Ophiolitic mélange separates ortho- and paratectonic Caledonides in western Ireland. Nature, 302, 50—52.
Ryan, P. D., M. D. Max and T. Kelly (1983): The petrochemistry of the basic volcanic rocks of the South Connemara Group (Ordovician), W. Ireland. Geol.Mag., 120, 141—152.
Sæther, T. and J. P. Nystuen (1981): Tectonic framework, stratigraphy, sedimentation and volcanism of the Late Precambrian Hedmark Group, Østerdalen, south Norway. Norsk geol.Tidsskr., 61, 193—211.
Sanzen-Baker, I. (1972): Stratigraphical relationships and sedimentary environments of the Silurian — early Old Red Sandstone of Pembrokeshire. Proc.Geol.Ass., 83, 139—164.
Sheppard, W. A. (1980): The ores and host rock geology of the Avoca Mines, Co. Wicklow, Ireland. Nor.geol.Unders., 360, 269—284.

Sjøstrand, T. (1978): Caledonian geology of the Kvarnbergsvattnet area, northern Jämtland, central Sweden. Sveriges Geol.Unders., C735, 107 pp.
Solyom, Z., R. Gorbatcher and I. Johansson (1979a): The Ottfjället Dolerites. Geochemistry of the dyke swarm in relation to the geodynamics of the Caledonide orogen in central Scandinavia. Sveriges Geol.Unders., C756, 38 pp.
Solyom, Z., P.G. Andreasson and I. Johansson (1979b): Geochemistry of amphibolites form Mt. Sylarna, Central Scandinavian Caledonides. Geol.Fören.Stockh.Förh., 101, 17–27.
Stephens, M.B. (1977): The Stekenjokk volcanites – a segment of a Lower Palaeozoic island arc complex. In: A. Bjørlykke, I. Lindahl and F.M. Vokes (Editors): Kaledonske Malmforekomster. BVLIS Tekniske Virksomhet, Trondheim, 24–36.
Stephens, M.B. (1980): Occurrence, nature and tectonic significance of volcanic and high-level intrusive rocks within the Swedish Caledonides. In: D.R. Wones (Editor): The Caledonides in the U.S.A. Virginia Polytechnic Inst. and State Univ. Geol.Sci., Mem. 2, 289–298.
Stephens, M.B. (1981): Evidence for Ordovician arc build-up and arc splitting in the Upper Allochthon of central Scandinavia. Terra cognita, 1, 75.
Stephens, M.B. (in press): Field relationships, petrochemistry and petrogenesis of the Stekenjokk volcanites, central Swedish Caledonides. Sveriges Geol.Unders., C786.
Stephens, M.B. and A. Senior (1981): The Norra Storfjället lens – an example of fore-arc basin sedimentation and volcanism in the Scandinavian Caledonides. Terra cognita, 1, 76–77.
Stephens, M.B., H. Furnes, B. Robins and B.A. Sturt (in press): Igneous activity within the Scandinavian Caledonides. In: D.G. Gee and B.A. Sturt (Editors): The Caledonide Orogen-Scandinavia and Related Areas. New York, John Wiley.
Stillman, C.J. (1981): Caledonian igneous activity. In: C.H. Holland (Editor): A geology of Ireland. Scottish Academic Press, Edinburgh, 83–106.
Stillman, C.J. and C.T. Williams (1979): Geochemistry and tectonic setting of some Upper Ordovician volcanic rocks in east and southeast Ireland. Earth Planet.Sci.Lett., 41, 288–310.
Stillman, C.J., K. Downes and E.J. Schiener (1974): Caradocian volcanic activity in east and southeast Ireland. Sci.Proc.R.Dubl.Soc., 5A, 87–98.
Strand, T. and O. Kulling (1972): Scandinavian Caledonides. Wiley-Interscience, London, 302 pp.
Sturt, B.A., I.R. Pringle and D.M. Ramsay (1978): The Finnmarkian phase of the Caledonian orogeny. J.geol.Soc.Lond., 135, 597–610.
Sturt, B.A., D. Roberts and H. Furnes (in press): A conspectus of Scandinavian Caledonian ophiolites. In: I.G. Gass, S.J. Lippard and A.W. Shelton (Editors): Ophiolites and oceanic lithosphere. Spec. publ. geol. Soc. Lond., 13, 381–391.
Thirlwall, M.F. (1981): Implications for Caledonian plate tectonic models of chemical data from volcanic rocks of the British Old Red Sandstone. J. geol. Soc. Lond., 138, 123–138.
Thirlwall, M.F. and B.J. Bluck (in press): Sr-Nd isotope and chemical evidence that the Ballantrae "ophiolite", S.W. Scotland, is polygenetic. Spec.Publ.Geol.Soc.Lond.
Thorpe, R.S. (1979): Late Precambrian igneous activity in southern Britain. In: A.L. Harris, C.H. Holland and B.E. Leake (Editors): The Caledonides of the British Isles – reviewed. Spec.Publ. Geol.Soc.Lond., 8, 579–584.
Trouw, R.A.J. (1973): Structural geology of the Marsfjällen area, Caledonides of Västerbotten, Sweden. Sveriges Geol.Unders., C689, 115 pp.
Van de Kamp, P.C. (1969): The Silurian volcanic rocks of the Mendip Hills, Somerset; and the Tortworth area, Gloucestershire, England. Geol.Mag., 106, 542–553.
Wilkinson, J.H. and J.R. Cann (1974): Trace elements and tectonic relationships of basaltic rocks in the Ballantræ igneous complex, Ayrshire. Geol.Mag., 111, 35–41.
Wolff, F.Chr. and D. Roberts (1980): Geology of the Trondheim region. Nor.geol.Unders., 356, 117–128.
Yardley, B.W.D., F.J. Vine and C.T. Baldwin (1982): The plate tectonic setting of NW Britain and Ireland in late Cambrian and early Ordovician times. J.geol.Soc.Lond., 139, 457–466.
Zachrisson, E. (1964): The Remdalen Syncline. Sveriges Geol.Unders., C596, 53 pp.
Zachrisson, E. (1966): A pillow lava locality in the Grong District, Norway. Norsk geol.Tidsskr., 46, 375–378.
Zachrisson, E. (1969): Caledonian geology of Northern Jämtland-Southern Västerbotten. Sveriges Geol.Unders., C644, 33 pp.
Zachrisson, E. (1973): The westerly extension of Seve rocks within the Seve-Köli Nappe Complex in the Scandinavian Caledonides. Geol.Fören.Stockh.Förh., 95, 243–251.
Ziegler, A.M., W.S. McKerrow, W.V. Burne and P. Baker (1969): Correlation and environmental setting of the Skomer Volcanic Group, Pembrokeshire. Proc.Geol.Ass., 80, 409–439.

Revised manuscript received 5 March 1984

The Role of Thrusting in the Scandinavian Caledonides

J. R. Hossack
Geology Department, City of London Polytechnic, Walburgh House, Bigland Street,
London E1 2NG, U.K.*

Key Words

thrust geometry
foreland
sole thrust
synorogenic surface balanced sections
tectonic shortening
continental collision

Abstract

The Scandinavian Caledonides are deeply eroded so that the original foreland and external thrust wedge are now only preserved in the Permian Oslo graben. Hence the foreland within the graben represents the only fixed pin-line for section balancing in the whole of Scandinavia. The sole thrust at the base of the thrust wedge follows the basal sediments just above the top of the underlying basement and dips at less than 2° towards the hinterland and a reconstructed synorogenic surface appears to dip at 4½° towards the foreland. Internally, the wedge is imbricated in the lower part but only folded in the upper part. The wedge initiated below sea-level in the mid Silurian but emerged by Wenlock times. Behind the wedge, the mountain belt consists of imbricated and thrusted basement and Proterozoic to Silurian cover sediments of the Baltic craton. These have typical thrust sheet geometry. Because the sole thrust cuts up section towards the foreland, older and older rocks are involved in the higher thrust sheets. Near the top of the tectonic pile, oceanic and Laurentian craton rocks have been obducted on top of the Baltic craton. The emplacement directions of the thrusts can be estimated from lineations, folds and lateral branch-lines and the amount of displacement estimated from abandonded horses and balanced sections. Over 400 km of shortening has occurred across southern Norway in the widest part of the Scandinavian Caledonides, mostly in the mid Silurian, at an average rate of 2.8 cm/yr.

---

* present adress: BP Petroleum Development Limited, Britannic House, Moor Lane, London EC2Y 9BU.

Introduction

The Caledonian chain stretches 1800 km through Scandinavia and has an exposed width of between 200 to 500 km (Fig. 1). This belt was formed by the tectonic emplacement of nappes and thrust sheets from the northwest up on to the Baltic craton at various times between the late Cambrian and the Silurian (Roberts & Sturt 1980). From balanced sections it can be estimated that there was at least 400 km of shortening across the widest part of the belt, between Oslo and the west coast, and this shortening produced a tectonic thickening of the continental crust from an initial thickness of 30 km to up to 70 km in the centre of the belt (Hossack & Cooper in press). Most of this orogenic contraction occurred by thrusting rather than folding so it is appropriate to review the geometry and origin of the thrust structures which play such an important role in the formation of the Scandinavian Caledonides.

Since the recognition of thrust faults in the 1880's there has always been two approaches to the study of thrust tectonics. One approach, largely developed in the Alps (Heim 1919, 1921, 1922), is to study the volume of rock which lies between major thrusts; to outline the stratigraphy and internal tectonic structures within this volume of rock and formally name the volume as a *nappe*. The other approach, pioneered by the geologists of the Geological Survey of Scotland (Peach et al. 1907), is to emphasize the geometry of the thrust faults and to name formally the individual thrusts from type localities. The volume of rock which exists above a named thrust in then called a *thrust sheet* and takes its name from the thrust beneath (Boyer & Elliott 1982). Applications of this second approach have been further developed in North America (Bally et al. 1966, Price & Mountjoy 1966, Dahlstrom 1970, Boyer & Elliott 1982).

Traditionally, the nappe approach has been favoured in Scandinavia (Gee 1975) but thrust geometry studies are now being attempted (Chapman et al. in press, Hossack et al. in press) and this review will concentrate on the latter approach. The review will use terminology which may not be familiar to most geologists but the interested reader can find descriptions of the geometry of thrusts in Dahlstrom 1970, Elliott & Johnson 1980, Butler 1982, and Boyer & Elliott 1982.

These two separate approaches are not antagonistic and should be regarded as parallel complimentary studies where there should be continuous cross-fertilization of ideas. For instance, the nappe approach leads to a description of lithotectonic stratigraphy and comparisons of adjacent nappes then may lead to speculations of palaeogeography and palinspastic restoration of the mountain belt. The second approach helps to clarify the geometry of thrust emplacement and its timing, how far the thrusts have moved and in what direction, in which order and the rate at which they have moved. By combining the two approaches, it should be feasible to carry out well-founded palinspastic restorations during different stages of the formation of the mountain belt.

Tectonic Sub-Divisions

The Scandinavian Caledonides can be separated into a series of parallel tectonic sub-divisions which from southeast to northwest have previously been named the autochthon, the parautochthon and the lower, middle and upper allochthon (Gee 1980). However a different classification scheme is used here which suggests a palaeogeographic origin for the thrust sheets of each division (Hossack & Cooper in press). The most external zone in the southeast is the *foreland* of Baltic craton. Thrusted on to the foreland, from the northwest, are a series of *cover thrust sheets*, largely composed of deformed sediments of Upper Proterozoic to Silurian age. Above these are a series of *crystalline thrust*

Thrusting in the Scandinavian Caledonides

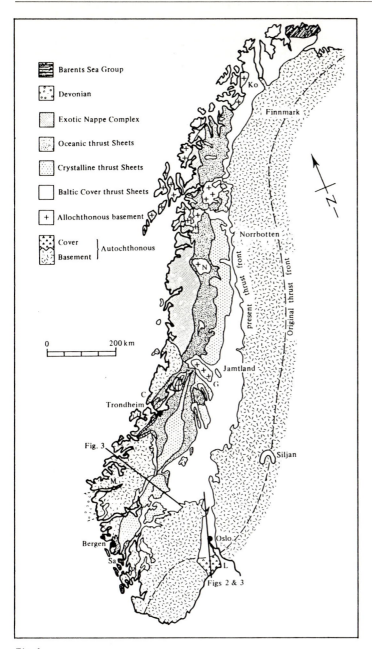

Fig. 1
Tectonic map of the Scandinavian Caledonides with the thrusted rocks named from their original palaeogeographic position of the thrust sheets prior to orogenesis (Hossack & Cooper in press). The present thrust front at Langesund (L) is indicated with a proposed original extension along strike which lies 150 km beyond the present thurst front.
T = Tømmerås, G = Grong, N = Nasafjäll,, K = Komagfjord, M = Møre. Lines of section of Figs. 2, 3 & 7 indicated.

sheets which consist of Precambrian basement rocks. Both the cover and the crystalline rocks are believed to have been derived from the Baltic craton. Above the latter sheets are the *oceanic thrust sheets* which are mainly composed of Lower Palaeozoic sediments and volcanics rocks which have formed on the edge of the Baltic craton or in back-arc basins or oceanic areas originally beyond the craton. Locally, within the thrust sheet pile are *windows* which expose Precambrian basement. The cores of these windows are traditionally regarded as autochthonous. However, more and more of these are being recognized as allochthonous (Kulling 1964, Thelander et al. 1980). These allochthonous windows are indicated on Fig. 1 but there is some doubt in my mind if the remaining windows figured as autochthonous are truly so. Finally, the highest sub-division consists of a group of fold nappes which Hossack & Cooper (in press) call the *exotic nappe complex*. These do not have the same geometry as the underlying thrust sheets and hence the term nappe has been retained.

The Foreland

This consists of Precambrian gneisses of Baltic shield with an overlying cover of Cambro-Silurian to lower Devonian sediments. The only area where this cover section is complete is in the south end of the Oslo graben at Langesund (Morley 1983) and in a few isolated areas in Sweden (Fig. 1). Elsewhere in the Scandinavian Caledonides, the cover sediments have been removed by later erosion so that the Precambrian shield forms the foreland. The Langesund section is the only area in Scandinavia which forms a fixed pin-line (Elliott & Johnson 1980) suitable for section balancing.

The Frontal Thrust Wedge

The sole thrust to the Scandinavian Caledonides is exposed as the frontal thrust on the northwest side of the Baltic shield foreland. However, the thrust is offset 150 km to the south within the Permian Oslo graben because of the downthrow in the graben centre (Fig. 1.) Unfortunately, the thrust front is not exposed in the graben because it is hidden by a younger Permian intrusion (Fig. 2) but the sole thrust is exposed within the graben. Here it forms a 150 km long *thrust flat* (Dahlstrom 1970) which lies within the lowest beds of the Cambro-Silurian section at the level of the Middle Cambrian black shales, barely 10–20 m above the underlying Precambrian basement. The imbricate thrusts, which are an important tectonic feature of the external thrust belt (Morley 1983) and form numerous tectonic repetitions of the Cambrian to middle Ordovician rocks, all splay off this basal detachment fault (Fig. 2). These thrusts are generally inclined towards the orogenic hinterland but many are inclined towards the foreland forming *triangle zones* and *pop-ups* (Elliott 1980). The imbricates are concentrated in the incompetent Cambrian to middle Ordovician beds, and die out upwards into *tip-lines* (Boyer & Elliott 1982). The upper Ordovician and Silurian beds are composed of thicker, more competent limestones which have merely folded into broad, open folds. The Oslo graben imbricate section is the most external part of the Scandinavian Caledonides still preserved and it is likely that similar sections once existed throughout the rest of Scandinavia in the area between the present and the proposed original thrust front (Fig. 1).

The composite section of Fig. 2 (Morley 1983) also includes a suggested synorogenic surface based on conodont alteration indices (CAI) (Bergstrom 1980, Hossack & Cooper in press). All the CAI in the Oslo graben have a value of 5 but these are believed to reflect the thermal alteration by the Permian intrusives and not the maximum depth

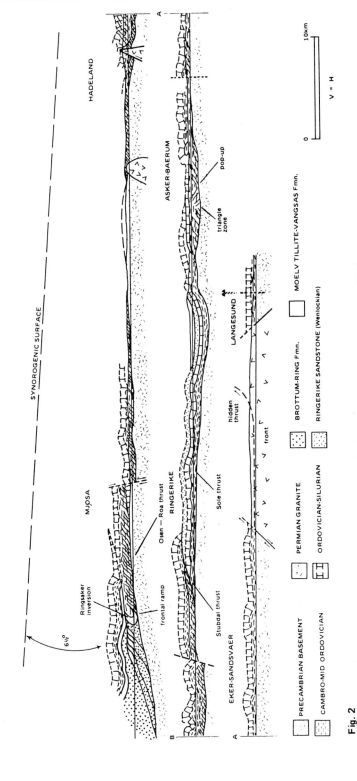

Fig. 2
Composite cross-section of the frontal thrust wedge in the Oslo graben (Morley 1983). The synorogenic surface is estimated from CAI in Sweden (Bergstrom 1980).

of burial. However, CAI are known from Jämtland to the northeast in Sweden and in the Swedish part of the foreland (Bergstrom op. cit.). At the thrust front, in Jämtland, the CAI have a value of 5 but fall off to values of 1–1½ at Siljan (Fig. 1) and the Baltic coast. This decrease can be interpreted as a reduction in the maximum depth of burial towards the southeast and the foreland. Assuming a geothermal gradient of 25°C/km, the position of the original surface which formed the maximum depth of burial i.e. the *synorogenic surface* in the deformed state (Boyer & Elliott 1982), can be estimated. The surface originally lay about 10 km above the present erosion level at Jämtland and had a slope of 4½° in the deformed state towards the foreland. In the same area, the sole thrust dips towards the hinterland at an angle of less than 2° (Gee et al. 1978). Hence the synorogenic surface and the sole thrust define a tapering wedge which thinned towards the foreland with a maximum internal taper of 6½°. Davis et al. (1983) have reviewed the importance of the wedge shape in thrust belts and using their graphs and equations it is possible to estimate that the Scandinavian frontal wedge formed with a Hubbert-Rubey fluid pressure ratio of $\lambda = 0.8$ and at 25°C/km had a brittle to ductile deformation transition at a syntectonic depth of 14 km at 200 MPa. The geometry of the Caledonian frontal wedge is remarkably close to that of the present-day Himalayas (Davis et al. op. cit.) Independent estimates (Hossack & Cooper in press) suggest that the thrusts in Scandinavia moved at a time averaged rate of 2.8 cm/yr.

The geometry of the Jämtland wedge taper has been projected along strike to the southwest into the plane of Fig. 2 to determine how much rock could be missing from the Oslo section. The suggested geometry indicates that most of the wedge is now missing. The highest part still remaining in the sucession is the Ringerike Sandstone of Wenlock age (Bockelie 1973). This unit is marine at the base but quickly passes up into continental beds. The sandstone can be interpreted as part of a clastic foredeep which was deposited ahead of the more internal moving thrusts and that the foredeep sediments themselves eventually became involved in the thrusting (Gee & Wilson 1974, Hossack et al. in press). Hence one way to account for the missing material within the thrust wedge is to postualte imbricated foredeep sediments all the way up to the missing synorogenic surface. Because the upper part of the Ringerike Sandstone is continental and the lower marine, the frontal thrust wedge was mainly subaerial throughout most of the tectonic history but was initiated as a submarine wedge.

The Oslo external wedge passes at its north end, at Mjøsa, into the Osen-Røa Nappe (Nystuen 1981) which forms one of the Baltic cover thrust sheets of Hossack & Cooper (in press). In the Osen-Røa thrust sheet, the sole thrust lies at the base of the Upper Proterozoic Brøttum Formation but cuts up section towards the Oslo external wedge across a *frontal ramp* (Elliott unpublished work, Butler 1982). It leaves the Brøttum Formation to cross the Ring Formation, the Ekre Shale and Vangsås Formation to reach the Middle Cambrian black shales which form the hanging- and foot-wall rocks of the Oslo external wedge (Morley 1983). The leading edge of the Mjøsa ramp also corresponds to the feather edge of onlapping Upper Proterozoic to Lower Palaeozoic sediments which originally thinned out against the Baltic shield foreland. Hence older and older sediments become involved in the thrusting towards the hinterland of the thrust belt. There is a geometric necessity to always form an anticline in the hanging-wall above a thrust ramp (Rich 1934) and in the Mjøsa area the Ringsaker inversion could have formed from the overturned and thrusted out limb of the hanging-wall ramp anticline (Fig. 2) (Suppe 1983). This frontal ramp structure extends in both directions along strike (Fig. 1) into the present thrust of the rest of the Caledonides. It is likely that this front corresponds elsewhere to a string of identical frontal ramps where the sole thrust has cut upwards

towards the foreland from the Upper Proterozoic to Palaeozoic sediments (Hossack & Cooper in press).
Most previous authors (e.g. Gee 1975) have separated a thin Cambro-Ordovician zone of sediments, which lies beneath this frontal ramp, from the thrust zone and maintained that the former represents part of the autochthonous foreland. However, in many cases these "autochthonous" rocks are folded, imbricated and cleaved and Hossack & Cooper (in press) have included this zone with the external thrust sheets. We speculate that this zone is frontal *duplex* (Boyer & Elliott 1982) which lies above the sole thrust which we place in the Middle Cambrian black shales just above the underlying Baltic shield gneisses.

The Cover, Crystalline and Oceanic Thrust Sheets

These terms refer to the imbricated and thrusted rocks which lie to the northwest of the Mjøsa frontal ramp and the conventional thrust front throughout the rest of Scandinavia. It is not possible to describe these rocks in detail in a review of this type but references to more detailed descriptions of rock type, structure, tectonic terminology and metamorphism are available in Andreasson & Gorbatschev (1980) and Hossack & Cooper (in press). The rocks above the thrust front were originally deposited as a conformable shelf sequence between the late Precambrian and the Silurian which onlapped and thinned towards the southeast on to the Baltic shield (Gee 1975). Locally, the thickness of the sediments varies between 3–5 km to the northwest of the frontal thrust but directly thins to 2 km within the external thrust wedge of the Oslo graben. As outlined above, the sole thrust cuts up section towards the foreland and dips backwards towards the hinterland. Hence older and older rocks appear within the hanging-walls of the higher thrust sheets. Eventually, Precambrian, basement rocks become involved in the thrusting. Locally, a few areas of basement occur in the cover thrust sheets but the sole finally cuts down towards the hinterland in restored geological cross-sections (c.f. Fig. 8) across frontal ramps into basement. On the hinterland side of these ramps are the crystalline thrust sheets of Fig. 1 (Hossack & Cooper in press). These crystalline sheets have been overthrust in turn by volcanic rocks and sediments from an area which originally lay to the west of the Baltic craton and these latter rocks make up the oceanic thrust sheets. Some of the volcanic rocks are ophiolitic (Furnes et al. 1982).
All these sheets have the geometry of *thrust sheets* (Boyer & Elliott 1982). They taper backwards to *trailing branch-lines* (Boyer & Elliott op. cit., Hossack 1983). Along strike, the sheets may wedge or taper into *oblique* or *lateral branch-lines* (Hossack op. cit.). Internally, the thrust sheets are imbricated by *emergent* or *duplex imbricates* (Boyer & Elliott op. cit.) and contain *frontal, lateral* and *oblique* ramps (Elliott unpubl. work, Butler 1982) and their attendant *ramp anticlines*. The tectonic style of thrust sheets is clearly distinct from fold nappes.
The thrust sheets of the Scandinavian Caledonides have two peculiar traits by which they appear to differ from previously described examples (e.g. Bally et al. 1966, Price & Mountjoy 1966, Dahlstrom 1970, Boyer & Elliott 1982). Firstly, the sheets are very long and thin in cross-section (Fig. 3b). I believe that the only way such long thin sheets could be emplaced was to be progressively accreted on to the foot-walls of the higher thrusts and the whole moved in piggy-back fashion towards the foreland (Bally et al. op. cit., Dahlstrom op. cit.).
Secondly, lens-like areas of thrusted rock exist which are completely separated from the main outcrop of the same thrust sheet. These have been previously described as mega-

boudins which became separated from their thrust sheets by a later period of crustal extension (Andreasson et al. 1978). However I interpret these as abandonded horses (Elliott unpub. work, Butler 1982), lenses of rock which have been left behind during thrusting (Hossack 1983).

The Exotic Nappe Complex

This is the highest tectonic unit now visible (Hossack & Cooper in press). Internally, the rocks are folded and refolded by tight similar folds which are separated by slide zones. Because the exotic nappe complex exhibits totally different deformation style, stretching directions, metamorphic grade and is not similar to any of the Scandinavian thrust units, we believe that these rocks represent the obducted leading edge of the North American Laurentian craton and that the oceanic thrust sheets beneath represent remnants of the Iapetus ocean which originally separated North America from the Baltic craton.

A Cross-Section Through the Caledonides

Detailed sections have been described by various authors (e.g. Gee 1975, 1978, Nystuen 1981, 1983, Chapman et al. in press, Hossack et al. in press). I shall concentrate in this review on the section from Oslo, through Valdres and Jotunheim, to the west coast of Norway (Hossack et al, op. cit.) (Fig. 3). But I shall then use the other detailed sections to pick out particular aspects of thrust tectonics in Scandinavia.

The longest cross-section through the Scandinavian Caledonides can be drawn from Langesund, northwards through the Oslo graben, to Mjøsa, and then northwestwards through the Jotunheim to the west coast (Figs. 1 & 3). The bend in the section is necessary to allow the local slip directions of the thrusts to remain in the plane of the section because only such sections can be properly restored to the undeformed state (Elliott 1983). The section begins in the external thrust wedge of the Oslo graben (Morley 1983) where the main thrust transport direction was towards the south. Only some of the more important tectonic structures can be shown at this scale. The sole thrust in the Cambrian shales, just above the basement, and the proposed synorogenic surfaces are indicated (Figs. 2 & 3). At Mjøsa the sole thrust ramps down section hindwards into the Upper Proterozoic sediments and forms the Ringsaker ramp anticline above. The section of Fig. 3b continues, with an offset to the southwest, through the Valdres area (Hossack et al. in press). Here there are three major external thrust sheets which are from the thrust front inwards the Aurdal duplex, the Synnfjell duplex and the Valdres thrust sheets respectively (Fig. 3). The sole thrust to these sheets can be traced from the thrust front over 50 km back underneath along the sides of the Aurdal valley. Imbricated late Precambrian to Ordovician sediments occur in the hanging-wall above the sole and autochthonous Middle Cambrian shales and Precambrian gneisses occur in the foot-wall beneath the sole. The most northern occurrence of the Cambrian shales is indicated in Fig. 3b. Here the sole thrust has a gentle dip of $1-2°$ towards the hinterland. The sole thrust re-appears within the external thrust sheets in the Vang and Beito windows (Hossack et al. in press) before disappearing beneath the crystalline thrust sheet of the Jotunheim. Only cover sediments are involved in the thrusting in the two frontal duplexes but in the Valdres thrust sheet Precambrian basement rocks make their first appearance. Hence the sole thrust has clearly cut up section from the hinterland towards the foreland.

The cover thrust sheets disappear to the northwest under the Jotun thrust sheet (Fig. 3) which is a large klippe that lies within a synformal depression and beyond is a large dome of Precambrian gneiss forming the Møre window which may or may not be autochthonous. If it is autochthonous, the sole thrust is domed up over the top of the window. If it is allochthonous and underlain by a *blind thrust* (Thompson 1981) then the thrust over the top of the window will no longer be the sole thrust. The Jotun thrust sheet can be traced around the southwest end of the Møre window to a trailing edge that disappears beneath the Devonian sedimentary beasins of west Norway (Hossack & Cooper in press, Hossack in press). Hence the Jotun thrust originally continued over the top of the window to the west coast (Fig. 3). The cover sheets beneath the Jotun thrust can be traced partly around the northeast side of the Møre window to a trailing branch-line which follows the long axis of the window and intersects the line of section (Fig. 3b).

There are volcanic rocks above the trailing edge of the Jotun thrust sheet on the west coast of Norway which have been correlated with the oceanic rocks of the Trondheim region (Skjerlie 1969). Hence it is likely that the oceanic thrust sheets originally covered the Jotun thrust sheet and this is indicated in Fig. 3.

The emplacement direction of the thrusts can be estimated using various tectonic structures such as slickenside striations, the symmetry axes of sheath folds, stretching lineations and lateral ramp structures and branch-lines (Hossack et al. in press). All the estimates in the Valdres part of the section are remarkably consistent giving a constant regional slip direction of 130° southeast. This contrasts with the southerly slip directions in the Oslo part of the section (Morley 1983). The section has been restored to its undeformed state after the manner of Dahlstrom (1969) in these two different slip directions (Fig. 8). Between Langesund and the trailing edge of the Jotun thrust there is over 400 km of shortening.

Out-Of-Sequence Thrusting

The structure of the Valdres section can be traced northeastwards along strike into the thrust structures of the Osen-Røa sheet (Nystuen 1981, 1983). Here imbricate thrusts with ramp and flat geometry are still present but one of the interesting features of this section is the possible occurrence of out-of-sequence thrusting (McClay & Coward 1981). It is generally held that thrust belts develop in a sequential piggy-back manner (Bally et al. 1966). The higher and more internal thrusts are believed to move first and then the more external thrusts develop and passively carry the higher thrusts on their backs. However, although this sequence is probably correct for most of the Scandinavian Caledonides, it is now apparent that locally thrusts stick during movement and then higher thrusts, which have already moved and then stopped, are forced to rejuvenate and move again over the top of lower thrusts. This rejuvenation may truncate fold and thrust structures in these lower thrusts giving the appearance that the main movement on the higher thrust post-dates the movement of the lower thrusts. In Nystuen's section (op. cit.) the Osen-Røa thrust is the lowest and has above it the Kvitvola thrust. The latter thrust may have moved in an out-of-sequence manner. At its leading edge (Fig. 4) the thrust clearly truncates foot-walls folds of the Osen-Røa thrust sheet. However, at its trailing edge the Kvitvola thrust is displaced and imbricated by thrusts climbing up out of the Osen-Røa thrust. Paradoxically, the Kvitvola thrust pre-dates and post-dates the Osen-Røa thrust. Out-of-sequence movement explains this paradox. Like the thrusts in Valdres (Figs. 3 & 8) large transport distances are clearly involved in Nystuen's sections where he estimates that the Osen-Røa thrust has moved between 200 and 400 km.

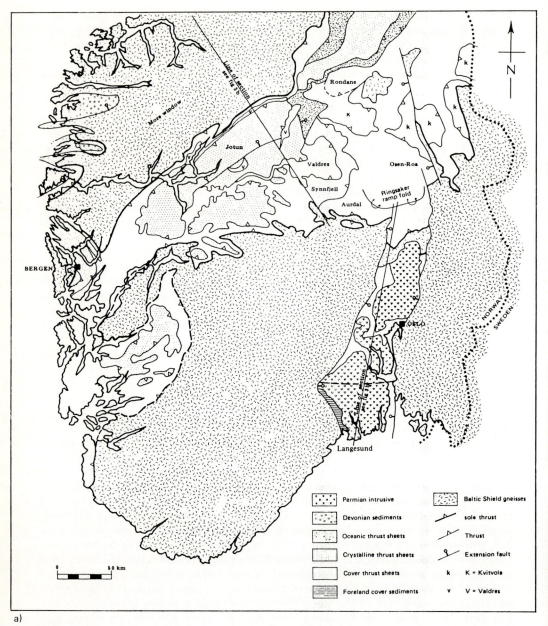

Fig. 3

a) Regional geological map of south Norway. b) Composite cross section through the Scandinavian Caledonides of south Norway from the foreland at Langesund, through the Oslo graben, and then through Valdres to the west coast. Line of section indicated in Fig. 3a. A = Aurdal duplex, S = Synnfjell duplex, V = Valdres thrust sheets. Ornamentation as in Fig. 3a.

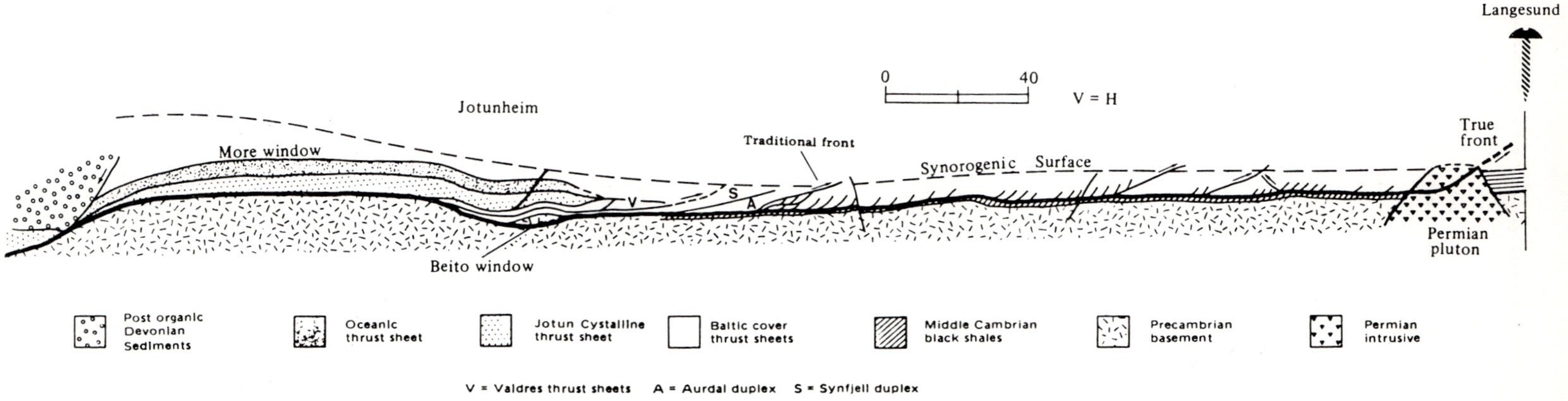
Langesund
0
40
V = H
Jotunheim
True
front
Traditional front
More window
Synorogenic Surface
V
S
A
Permian
pluton
Beito window
Post organic
Devonian
Sediments
Oceanic
thrust sheet
Jotun Cystalline
thrust sheet
Baltic cover
thrust sheets
Middle Cambrian
black shales
Precambrian
basement
Permian
intrusive
V = Valdres thrust sheets
A = Aurdal duplex
S = Synfjell duplex

b)

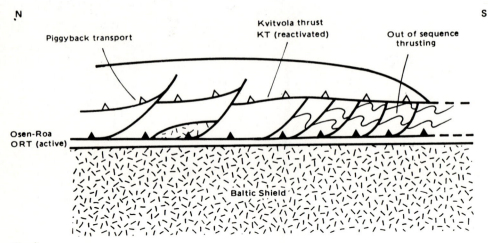

Fig. 4
Sketch cross-section through the Osen-Røa and Kvitvola thrusts (Nystuen 1983) showing potential out-of-sequence thrusting at the leading edge and piggy-back thrusting at the trailing edge of the Kvitvola thrust.

At their trailing edges, the Osen-Røa and Kvitvola thrust sheets taper into trailing branch-lines beneath the higher thrusts of the Trondheim Nappe (Roberts & Wolff 1981) which is composed of crystalline and oceanic thrust sheets (Fig. 1). These in turn disappear down-dip beneath the exotic nappe complex.

Branch-Line Patterns

The branch-line patterns of the various thrust sheets in the Trondheim area (Fig. 1) illustrate another important facet of the Scandinavian Caledonides (Hossack 1983). Two of these sheets, the Tännas-Särv and the Seve have characteristic trailing branch-lines composed of salients and re-entrants (Fig. 5). These have opposed parallel sides which are interpreted as *lateral ramps* (Elliott unpubl. work, Butler 1982) which originally had mirror-image geometry that faced one another. Only one slip direction is geometrically possible which will produce orogenic contraction, namely parallel to the two sides of the re-entrant (Elliott & Johnson 1980). Any other slip direction will cross both sides obliquely and produce a contraction fault on one side and an extension fault on the other. The lateral branch-lines probably outline the slip direction of the thrusts and in the case of the Tännas-Särv and the Seve thrusts the slip vectors were between 155 and 165° southeast. However, in much of this area the prominent stretching lineation is towards 100—130° southeast and this could also be interpreted as thrust slip directions. I believe that both directions are significant and each represents different stages of the thrust movement. However I am not certain yet which was the earlier stage.

The branch-line pattern of the Trondheim area (Hossack 1983) also illustrates another important property of thrusts, namely that isolated *horses* (Elliott & Johnson 1980) of a particular thrust sheet may be left trailing behind the thrust during movement. In the case of the Köli thrust (Hossack op. cit. fig. 8d) there is an abandonded horse 50 km behind the trailing branch-line of the same thrust. Hence this is an estimate of the minimum distance of displacement on this thrust. In the case of the Särv thrust (Fig. 5)

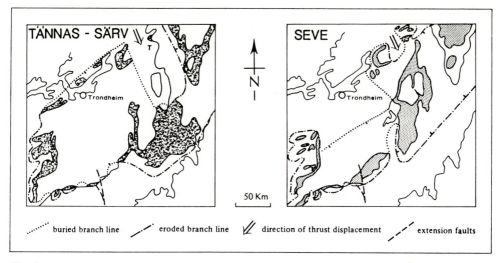

Fig. 5
Branch-line patterns in the Tännas-Särv and Seve thrusts in the Trondheim area. Re-entrants represent lateral ramps which may lie in the thrust slip direction between 155° and 165° southeast. An abandoned horse at Turtbakktjørna (T) suggests a minimum slip distance of 100 km on the Särv thrust.

there is an abandonded horse 100 km behind the trailing branch-line. This is the minimum displacement for this latter thrust.

The trailing branch-lines of the Scandinavian Caledonides can be extremely long. Zachrisson (1973) figured the tapering western edge or the trailing branch-line of the Seve thrust (Fig. 6). This structure stretches over 800 km along the strike of the mountain belt. Because branch-lines are initiated from ramp structures, there must have been a frontal ramp of equivalent dimensions present during the formation of the Seve thrust (Hossack 1983).

Allochthonous Windows

Another significant thrust structure is displayed in the segment northeast of the Grong window (Fig. 1) and is the allochthonous core to the Nasafjäll window (Thelander et al. 1980) (Fig. 6). The Precambrian basement in the core of this window was previously considered to be autochthonous but Thelander et al. (op. cit.) describe the presence of three inner windows which are framed by Cambro-Ordovician sediments with inner cores of gneisses. They interpret the main window as a unit of Precambrian volcanic rocks which has overthrust a lower unit of gneiss and Cambro-Ordovician cover. They speculate that other windows in the region have allochthonous cores. Similarly, Chapman et al. (in press) suggest that the Komagfjord window in north Norway is composed of a horse of thrusted basement in the core which is underlain by a thrust. All these allochthonous windows are indicated in Fig. 1. I have included in the allochthonous windows the Tømmerås window of central Norway where I have personally observed that the basement in the centre of the window is imbricated with the overlying Cambro-Ordovician sediments and forms a duplex window (Boyer & Elliott 1982). Hence the inner core has to be allochthonous. Similarly, I speculate that the adjacent Grong window

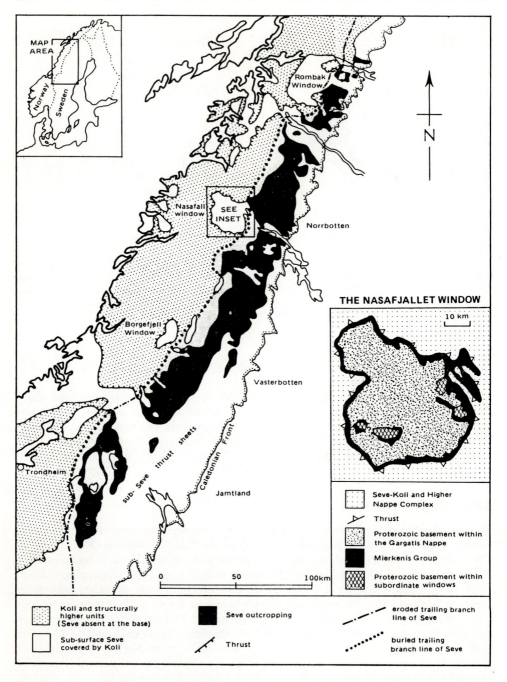

Fig. 6
The trailing branch-line of the Seve thrust stretching 800 km along the strike of the Scandinavian Caledonides (Zachrisson 1973). Inset map the structure of the Nasafjäll window (Thelander et al. 1980).

has allochthonous basement in the core. The "autochthonous" windows of south Norway are shaded differently (Fig. 1) but I believe that it is likely that they are also underlain by thrust faults.

Balanced Sections

These sections (Dahlstrom 1969) have only been drawn up for a few areas in the Scandinavian Caledonides (Chapman et al. in press, Hossack et al. in press, Morley 1983). A *viable* geological cross-section (Elliott 1983) has to be restored to the undeformed state so that there are no holes or gaps left in the restored section. It is also necessary to draw tectonic structures which exist in the area which follow clearly stated tectonic rules. Such a section is *admissible* and by definition a *balanced* cross-section is both viable and admissible (Elliott op. cit.).

Chapman et al. (in press) have drawn a series of cross-sections through the external cover thrusts of Finnmark, north Norway (Figs. 1 & 7). The latter figure is an example of a line-length balanced section (Hossack 1979). Both foreland- and hinterland-dipping thrusts are present and all cut up section in the direction of transport. Listric shapes are common and some of the hanging- and foot-wall cutoffs can be matched up. All the faults join a common sole thrust at depth and the section has undergone 25 km shortening (equivalent to -29% natural strain). Chapman et al. (op. cit.) have measured a maximum of -63% natural strain shortening in their area. The contraction occurred at various times between the late Cambrian and middle Silurian at a time averaged rate of 0.75 cm/yr.

In areas where the rocks have a cleavage or have been folded by similar folds it is necessary to restore by area balancing (Hossack 1979, Elliott & Johnson 1980). Briefly, the section can be restored if the deformed area can be divided by the original *unstrained* stratigraphic thickness of the section. An example of this kind of balancing is illustrated by the Valdres thrust sheets in Fig. 8. This cross-section represents a restored version of Fig. 3 (Hossack & Cooper in press). The section between Langesund and the Ringsaker hanging-wall ramp anticline has been restored by line-length balancing (Morley 1983). The section north of here, including the Aurdal and Synnfjell duplexes, and the Valdres thrust sheets, has been restored by area balancing. For instance, although many of the rocks in the latter sheets are highly deformed and strongly cleaved and finite strain estimates can be made from deformed conglomerates (Hossack 1968, 1978), there is one section of undeformed rock at Mellane (Loeschke & Nickelsen 1968). The stratigraphic thickness of 3 km in this undeformed section has been used to unstrain the whole of the Valdres thrust sheets by area balancing. A comparison of the initial and final lengths of the Valdres sheets allows an independent estimate of the internal ductile elongation to be made. The average cleavage-forming elongation of the thrust sheets is estimated to be $+109\%$ natural strain in a northwest-southeast direction. This is remarkably close to the regional average elongation of $+96\%$ natural strain measured in the deformed conglomerates (Hossack 1978). This is a difference of only 13% and suggests that area balancing is a simple but highly effective way of restoring mountain belts. The remainder of the section northwest of the Valdres sheets has been restored by line-length balancing. The section has a minimum shortening of 400 km or -69% natural strain. Included in the line of the section is an estimate of the minimum amount of oceanic crust which has been obducted on to the Baltic shield in the Trondheim section (Hossack 1983, Hossack & Cooper in press). The depth to Moho has been estimated from the geophysical data of Sellevoll (1973). Most of the 400 km shortening has occurred in the mid-Silurian phase of orogenesis at an average thrusting rate of 2.8 cm/yr (Hossack & Cooper op. cit.).

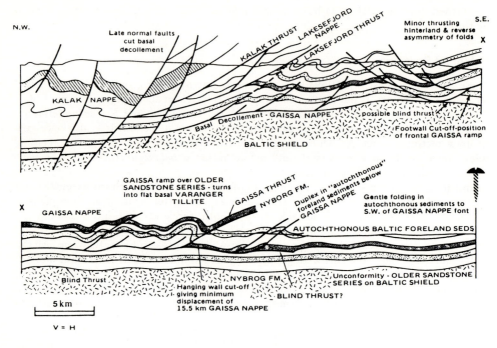

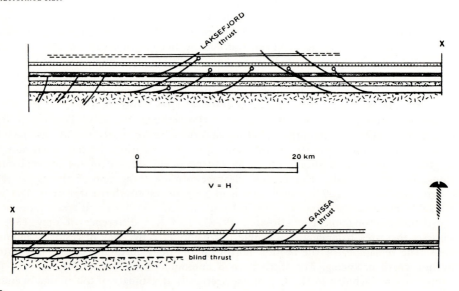

Fig. 7
Balanced bed-length section through the Finnmark region (Chapman *et al.* in press) a) deformed section b) undeformed section prepared by the present author. Note the reduction in scale of the restored section. V = H.

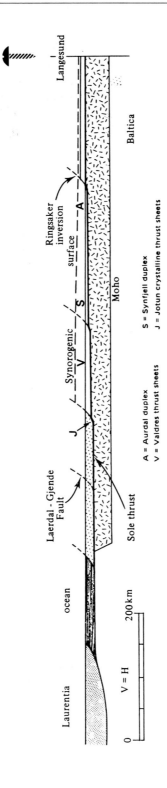

Fig. 8
Balanced section through the Scandinavian Caledonides based on the section of Fig. 3. The section from the pin-line at Langesund up to Ringsaker has been line-length balanced; the section from Ringsaker to the Jotun thrust has been area balanced (Hossack 1979). A = Aurdal duplex, S = Synnfjell duplex, V = Valdres thrust sheets, J = Jotun thrust sheet, black = oceanic crust. Depth to Moho estimated from the geophysical data of Sellevoll (1973).

Conclusions

The Scandinavian Caledonides are an example of a deeply eroded thrust belt. A shelf series of Upper Proterozoic to Silurian sediments, which originally lay on the edge of the Baltic craton, has been moved towards the southeast and telescoped by thrusting during continental collision between the Baltic and Laurentian cratons. The imbricated Baltic cover sediments are overlain by thrusted Baltic crystalline rocks and oceanic volcanics and sediments and finally at the top of the structural succession by continental rocks which are believed to represent the leading edge of the Laurentian craton. The sole thrust to the whole belt can be traced from the foreland at Langesund over 400 km into the interior of the belt. Near the margin of the orogenic belt, the sole thrust lies in Middle Cambrian black shales which immediately overlie the gneisses of the Baltic shield. Hinterland- and foreland-dipping thrusts climb up out of this sole thrust and produce numerous thrust imbricate repetitions of the overlying Cambro-Ordovician. The thrusts die out upwards into the folded, more competent Silurian rocks at the top of the preserved section. A missing group of foredeep sediments might have existed above the present stratigraphy all the way up to a postulated synorogenic surface which probably had a 4½° dip in the deformed state towards the foreland. The sole thrust to this external wedge dips now at less than 2° towards the hinterland so that the thrust wedge in the deformed state had an internal taper of 6½° towards the foreland. This tapering wedge is remarkably similar to the present external zone of the Himalayas, another thrust belt produced by continental collision. The only foreland cover rocks still exposed beyond the external thrust wedge are flat-lying, unfolded and uncleaved rocks at the south end of the Oslo graben at Langesund. This foreland and external thrust wedge are preserved in the graben by the Permian downthrow in the centre. They have been removed by later erosion from the rest of the Scandinavian thrust belt. The external thrust wedge was initiated in the mid Silurian as a submarine wedge but by Wenlock times had become a subaerial thrust wedge.

At Mjøsa the sole thrust at the base of the external wedge cuts down section backwards into older Proterozoic sediments across a frontal ramp which has the Ringsaker hanging-wall anticline above. This ramp is in the position of the present thrust front in the rest of Scandinavia. In the foot-wall beneath the front the sole thrust still follows the Cambrian shales just above the basement.

The thrusted rocks behind the thrust front all have the geometry of classic thrust sheets. They taper backwards and along strike into branch-lines. Internally, they are imbricated by listric and stepped thrusts which form emergent fans or duplexes. These thrusts cut stratigraphically deeper towards the orogenic hinterland and eventually involve Precambrian gneisses of the Baltic shield in the deformation. Hence the Baltic shield rocks are thrusted as lenses within windows or form major thrust sheets which overlie the cover sheets. The emplacement directions of individual thrusts can be estimated using lineations, sheath folds and branch-lines and the amount of displacement can be estimated using the position of abandonded horses of a particular thrust sheet or by constructing balanced cross-sections. Orogenic contraction varies locally from 29 to 69% natural strain and displacements of over 400 km can be estimated in a balanced section that completely crosses the exposed thrust belt. Some of the displacement on the higher thrusts occurred in the late Cambrian and early to mid Ordovician but the main thrusting occurred between the mid and late Silurian with a time average rate of thrust movement of 2.8 cm/yr.

Acknowledgements

I would like to thank Jane Gilotti, Chris Morley and Mark Cooper for their help with this manuscript. I would also like to thank Alan Sutton for drawing the illustrations.

References

Andreasson, P.-G., Gee, D.G. and Kumpulainen, R. (1978): Some remarks on the Steinkjer megaboudin. Norsk.geol.Tidsskr. 58, 305–307.
Andreasson, P.-G. and Gorbatschev, R. (1980): Metamorphism in extensive nappe terrains: a study of the Central Scandinavian Caledonides. Geol. Fören. Stockholm Förh. 102, 335–357.
Bally, A.W., Gordy, P.L. and Stewart, G.A. (1966): Structure, seismic data, and orogenic evolution of southern Canadian Rocky Mountains. Bull. Can. Petrol. Geol. 14, 337–81.
Bergstrom, S.M. (1980): Conodonts as palaeotemperature tools in Ordovician rocks of the Caledonides and adjacent areas in Scandinavia and the British Isles. Geol.för.Stock. Förh. 102, 377–92.
Bockelie, J.F. (1973): The presence of Prunocystites (Cystoiden) in Stage 9a of Ringerike, Norway. Nor. Geol. Tidskr. 53, 317–21.
Boyer, S.E. and Elliott, D. (1982): Thrust systems Bull. Am Assoc. Petrol. Geol. 66, 1196–230.
Butler, R.W.H. (1982): The terminology of structures in thrust belts. J. Struct. Geol. 4, 239–45.
Chapman, T.J., Gayer, R.A. and Williams, G.D. (in press): Structural cross-sections through the Finnmark Caledonides and timing of the Finnmark event. In: Gee, D.G. and Sturt, B.A. (eds.) "The Caledonide orogen-Scandinavia and related areas". J. Wiley & Sons, New York.
Dahlstrom, C.D.A. (1969): Balanced cross-sections. Can. J. Earth Sci. 6, 743–57.
Dahlstrom, C.D.A. (1970): Structural geology in the eastern margin of the Canadian Rocky Mountains. Bull. Can. Petrol. Geol. 18, 332–406.
Davis, D., Suppe, J. and Dahlen, F.A. (1983): Mechanics of fold-and-thrust belts and accretionary wedges. J. Geophys. Res. 88, 1153–72.
Elliott, D. and Johnson, M.R.W. (1980): The structural evolution of the northern part of the Moine thrust zone. Trans. R. Soc. Edinb. Earth Sci. 71, 69–96.
Elliott, D. (1980): How do thrust belts form? Bull. Am. Assoc. Petrol. Geol. 64/65, 704. Abstract Ann. Meeting Denver.
Elliott, D. (1983): The construction of balanced cross-sections. J. Struct. Geol. 5, 101.
Furnes, H., Thon, A, Nordås, J. and Garman, L.B. (1982): Geochemistry of Caledonian metabasalts from some Norwegian ophiolite fragments. Contrib. Mineral. Petrol. 79, 295–307.
Gee, D.G. (1975): A tectonic model for the central part of the Scandinavian Caledonides. Am J. Sci. 275, 468–515.
Gee, D.G. and Wilson, M.R (1974): The age of orogenic deformation in the Swedish Caledonides. Am. J. Sci. 274, 1–9.
Gee, D.G., Kumpulainen, R. and Thelander, T. (1978): The Tåsjön decollement, central Swedish Caledonides. Sver. geol. Unders. C Nr 742, 1–35.
Gee, D.G. (1980): Basement-cover relationships in the central Scandinavian Caledonides. Geol. för. Stock. Förh. 102, 455–474.
Heim, A. (1919, 1921, 1922): Geologie der Schweiz. 3 vols. Tauschnitz, Leipzig.
Hossack, J.R. (1968): Pebble deformation and thrusting in the Bygdin area (southern Norway). Tectonophysics 5, 315–39.
Hossack, J.R. (1978): The correction of stratigraphic sections for tectonic finite strain in the Bygdin area, Norway. J. geol. Soc. Lond. 135, 229–41.
Hossack, J.R. (1979): The use of balanced cross-sections in the calculation of orogenic contraction: A review. J. geol. Soc. Lond. 136, 705–11.
Hossack, J.R. (1983): A cross-section through the Scandinavian Caledonides constructed with the aid of branch-line maps. J. Struct. Geol. 5, 103–11.
Hossack, J.R. (in press): The geometry of listric growth faults in the Devonian basins of Sunnfjord, west Norway. J. geol. Soc. Lond.
Hossack, J.R. and Cooper, M.A. (in press): Collision tectonics in the Scandinavian Caledonides. J. geo. Soc. Lond.
Hossack, J.R., Garton, M.R. and Nickelsen, R.P. (in press): The geological section from the foreland up to the Jotun thrust sheet in the Valdres area, south Norway. In: Gee, D.G. and Sturt, B.A. (eds.): "The Caledonide orogen-Scandinavia and related areas". J. Wiley, New York.

Kulling, O. (1964): Oversikt over Norra Norbottens fjällens Kaledonberggrund. Sver. geol. Unders. Ba Nr. 19, 1–166.

Loeschke, J. and Nickelsen, R. P. (1968): On the age and positions of the Valdres Sparagmite in Slidre (Southern Norway). N. Jb. Geol. Palänt. Abh. 131, 337–67.

McClay, K. R. and Coward, M. P. (1981): The Moine Thrust Zone: an overview. In: McClay, K. R. and Price, N. J. (eds.): "Thrust and Nappe Tectonics". Geol. Soc. Lond. Spec. Pub. No 9, 241–260.

Morley, C. K. (1983): The structural geology of the southern Caledonides, South Norway, between Langesund and Mjøsa. Unpubl. Ph.D. Thesis. City of London Polytechnic.

Nystuen, J.-P. (1981): The late Precambrian "Sparagmites" of southern Norway: a major Caledonian allochthon- the Osen-Røa Nappe Complex. Am J. Sci. 281, 69–94.

Nystuen, J.-P. (1983): Nappe and thrust structures in the Sparagmite region, southern Norway. Nor. geol. Unders. 380.

Peach, B. N., Horne, J., Gunn, W., Clough, C. T. and Hinxman, L. W. (1907): The geological structure of the north-west Highlands of Scotland. Mem. Geol. Surv. G.B.

Price, R. A. and Mountjoy, E. W. (1970): Geological structure of the Canadian Rocky Mountains between Bow and Athabasca rivers- a progress report. In: Wheeler, J. O. (ed.): "Structure of the southern Canadian Cordillera". Spec. Pap. geol. Assoc. Can. 6, 7–25.

Rich, J. L. (1934): Mechanics of low angle overthrust faulting illustrated by Cumberland thrust block, Virginia, Kentucky and Tennessee. Bull. Am. Assoc. Petrol. Geol. 18, 1534–96.

Roberts, D. and Sturt, B. A. (1980): Caledonian deformation in Norway. J. geol. Soc. Lond. 137, 241–51.

Roberts, D. and Wolff, F.-C. (1981): Tectonostratigraphic development of the Trondheim region Caledonides, Central Norway. J. Struct. Geol. 3, 487–94.

Sellevoll, M. A. (1973): Mohorovicic discontinuity beneath Fennoscandia and adjacent parts of the Norwegian Sea and the North Sea. Tectonophysics 20, 359–366.

Suppe, J. (1983): Geometry and kinematics of fault-bend folding. Am. J. Sci. 283, 684–721.

Thelander, T., Bakker, E. and Nicholson, R. (1980): Basement cover relationships in the Nasafjället window, central Swedish Caledonides. Geol. för Stock. Förh. 102, 569–80.

Thompson, R. I. (1981): The nature and significance of large blind thrusts within the northern Rocky Mountains of Canada. In: McClay, R. and Price, N. J. (eds.): "Thrust and nappe tectonics". Geol. Soc. Lond. Spec. Pub. No 9, 449–62.

Skjerlie, F. J. (1969): The pre-Devonian rocks in the Askvoll-Goular area and adjacent districts, western Norway. Nor. geol. Unders. 258, 325–59.

Zachrisson, E. (1973): The westerly extension of Seve rocks within the Seve-Köli Nappe Complex in the Scandinavian Caledonides. Geol. för. Stock. Förh. 95, 243–51.

Revised manuscript received 9 Jan. 1984

A Major Stratigraphical and Metamorphic Inversion in the Upper Allochthon of the Scandinavian Caledonides

R. Mason
Department of Geology, University College, Gower Street, London WC1E 6BT, U.K.

Key Words

Scandinavian Caledonides
stratigraphic inversion
metamorphic inversion
ophiolite complex

Abstract

There is a large-scale stratigraphical and metamorphic inversion in the Upper Allochthon of the Scandinavian Caledonides along the Bodø-Kvikkjokk cross-section just north of the Arctic Circle. The inversion is genetically associated with the obduction and progressive deformation of an undisrupted ophiolite fragment in which the gabbro layer has been intruded into rocks showing pre-Scandian deformation and metamorphism. It is suggested that inversion may be widespread in the Upper Allochthon, which may therefore consist of fold-nappe rather than thrust-sheet units.

Introduction

The Caledonian orogenic belt of Scandinavia is a classic overthrust terrain, in which only the south-eastern strip of the belt, varying in width from 140 km to 300 km, is present (Gayer 1973, Nicholson 1974, Binns 1978). There is a major thrust-belt, running from SW to NE, bringing thrust-sheets of parautochthonous and allochthonous rocks of Proterozoic and Lower Palaeozoic age above the Baltic Shield and a thin cover of autochthonous sediments (Kulling & Strand 1972, Gee & Zachrisson 1979). The recognition of basalts with oceanic geochemical affinities and of fragments of ophiolites in the allochthon of the Scandinavian Caledonides has added a new dimension to the discussion of the tectonics (Gale & Roberts 1974, Furnes et al. 1980). One such ophiolite fragment is in the copper-mining district of Sulitjelma, Nordland Province, Norway (67°N, 16°E) (Boyle 1980). Because this district is well-known outside Scandinavia as a type area for progressive regional metamorphism (Vogt 1927, Mason 1978, Turner 1981), gabbro petrology (Mason 1971) and strata-bound sulphide mineralisation (Wilson 1973), I would like to draw attention to the implications of the discovery of the ophiolite in these fields and also for the wider tectonic interpretation of the Scandinavian Caledonides.

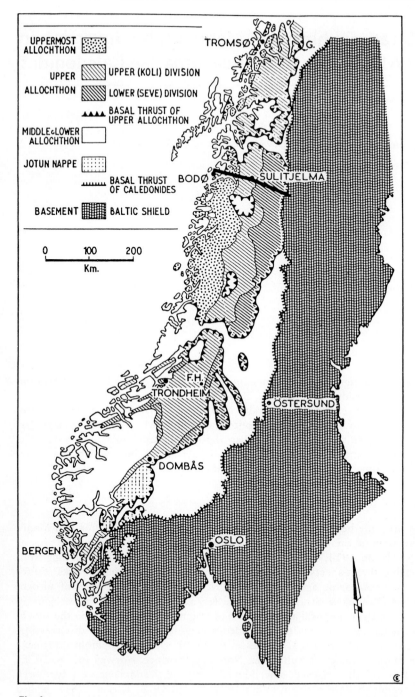

Fig. 1
Outline tectonic map of the Scandinavian Caledonides, after Roberts et al. (1981). The line of the section in Fig. 2 is indicated. F.H. — position of the Fongen-Hyllingen intrusion, G. — Guolasjav'ri (see text).

Table 1: Structural and Lithostratigraphical Sequences in the Bodø — Kvikkjokk Cross-section.
(The arrow shows the part of the sequence which is thought to be continuous, and inverted)

Major tectonic units	Tectonic subdivisions	Lithostratigraphical units	Informal names or predominant rock-types
Uppermost Allochthon	Fauske Nappe		Fauske marble schists conglomerates
Upper Allochthon	Gasak Nappe	Skaiti Supergroup	schists amphibolites marbles psammites gneisses granite intrusions
	Köli Division	Sulitjelma ophiolite complex	Sulitjelma gabbro Sulitjelma amphibolites
		Furulund Group	Furulund schist
		Sjønstå Group	Sjønstå gneiss Muorki schist
	Seve Division	Pieske Group	Pieske Marble
Lower or Middle Allochthon			Kvikkjokk mylonite complex

The Bodø-Kvikkjokk Cross-Section

Sulitjelma lies in a region of plunge depression between the culminations of Nasafjell to the SW and Tysfjord to the NE (Fig. 1). Most of the rocks of the district belong to the Upper Allochthon of the Central Scandinavian Caledonides (Kulling 1972, Gee & Zachrisson 1979). They are metamorphosed sedimentary and volcanic rocks intruded by gabbros and granites. The volcanic and sedimentary rocks show a well-defined lithostratigraphic sequence (Table 1) (Nicholson & Rutland 1969). Fossils in the lower part of the Furulund Group indicate a late Ordovician or early Silurian age (Vogt 1927, Gee & Wilson 1974). The copper ores occur close to or at the boundary between the Furulund Group and the Sulitjelma amphibolites. There is sulphide mineralisation at this level throughout the district, with local concentrations of massive and disseminated pyrite, chalcopyrite and sphalerite, forming economic ore-bodies (Wilson 1973). The Furulund Group shows a sequence of metamorphic zones of Barrovian type, of particular interest because the isograds cut across the lithostratigraphical units (Mason 1978). The mountains on the Norwegian-Swedish international boundary are of gabbro, showing rhythmic and cryptic layering.

The Inversion in the Upper Allochthon

The Sulitjelma amphibolites and the gabbro have been shown by Boyle (1980) to be members of an undisrupted but deformed ophiolite. The sheeted dyke complex is up to 1.5 km thick. The ophiolite is predominantly upside-down, on the evidence of the arrangement of its different members and of pillow structures and relationships between pillows and overlying sediments (Boyle et al. 1979). The major and trace element compositions of the lavas, sheeted dykes and gabbros show that they belong to one magmatic suite of oceanic type (Boyle et al. 1984). This implies that the sulphide ores are of the volcanic-exhalative Cyprus type, formed on ocean floor. The Furulund Group is stratigraphically inverted throughout, and its earliest folds are downward-facing (Kirk & Mason 1984).

The stratigraphical inversion is accompanied by inversion of the metamorphic zones, with higher grade rocks lying above lower grade (Mason 1984). The existence of this metamorphic inversion has been demonstrated by mapping the garnet isograd underground in a hydroelectric tunnel. The isograd cuts across the lithostratigraphic units and major fold axial surfaces, and across various thrust horizons which have been proposed in the area (Henley 1970). Although the gabbro shows magmatic contacts with the sheeted complex in some places (Mason 1971, Boyle 1980) in others it shows unambiguous intrusive relationships with the Skaiti Supergroup of the Upper Allochthon, which includes amphibolites, marbles, schists and conglomerates. In the contact aureole, and in xenoliths, relict folds and boudins are preserved and regional metamorphic minerals are seen partially or completely replaced by contact metamorphic minerals, for example kyanite by sillimanite, and garnet by orthopyroxene, cordierite and spinel (Mason 1971, 1980). Thus the Skaiti Supergroup must have undergone deformation and metamorphism before the construction of the ophiolite and the deposition of the Furulund Group (Table 2). It was presumably pre-existing continental crust on one side of the oceanic rift represented by the ophiolite. The only events in Table 2 which have been dated are the deposition of the Furulund Group and its metamorphism (Wilson 1971). The dating of the intrusion of the Sulitjelma gabbro by assuming that it was contemporary with the intrusion of the Furulund granite is wrong (Mason 1981). The metamorphism of the Furulund Group belongs to the Scandian event of 400–440 Ma B.P., until now considered to be the only significant Caledonian event in this part of the Caledonides (Gee & Wilson 1974, Roberts & Sturt 1980). The earlier deformation and metamorphism of the Skaiti Supergroup may correlate with the early Caledonian Finnmarkian event of northern Norway, or may be Precambrian. Thus at Sulitjelma there is a major stratigraphical inversion at the top of the Upper Allochthon, with the highest rocks in the structural sequence showing evidence of older deformation and metamorphism than those below. The upper part of the Skaiti Supergroup is right way up (Boyle et al. 1984) and it is therefore thought that the inverted succession of the ophiolite and Furulund Group is the inverted limb of a major anticlinal fold (Eig. 2). This structure suggests that the upper part of the Upper Allochthon in the Bodø-Kvikkjokk section is a major fold, rather than a thrust-sheet nappe.

Fig. 3 summarises the sequence of events in the Sulitjelma part of the Bodø-Kvikkjokk section (Fig. 2). The imposition of the metamorphic isograds on the pre-existing major folds is demonstrated by the downward facing of the cleavage in the Furulund Group (Kirk & Mason 1984) and by the overgrowth of early fabrics in the garnet porphroblasts (Henley 1970, Mason 1978). I suggest that the obduction of the ophiolite, the inversion of the stratigraphical sequence and the imposition of the inverted isograds represent successive stages in a continuous progressive deformational process (Mason 1984).

Stratigraphical and Metamorphic Inversion — Scandinavia

Table 2: Time sequence of depositional, deformational and petrological events in the Bodø — Kvikkjokk Cross-section.

Event name	Structural and depositional events	Petrological events
Scandian (late)	Open major and minor folding of schistosity	Greenschist facies retrograde matamorphism Thermal relaxation from Scandian peak, fixing of metamorphic assemblages, Rb-Sr and K-Ar ages 400 – 440 Ma B.P.
(early)	Tight major and minor folding, development of main schistosity Onset of deformation of Furulund Group and nappe transport	Peak of metamorphism
Rift closing phase	Deformation of parts of ophiolite cut by latest dykes of sheeted complex	Foliation developed in deformed gabbro
Rift opening phase	Construction of main part of ophiolite, deposition of Furulund Group	Intrusion of Sulitjelma gabbro, contact metamorphism of Skaiti Supergroup
Pre-rifting	Deformation of Skaiti Supergroup	Barrovian metamorphism of Skaiti Supergroup
	Deposition of Skaiti Supergroup	

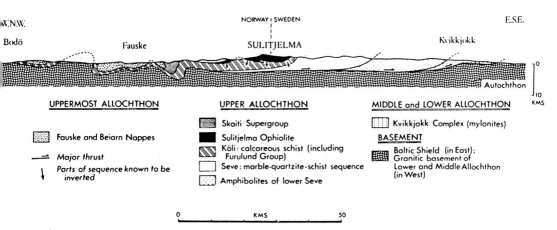

Fig. 2
Cross-section through the Scandinavian Caledonides from Bodø on the Norwegian coast, to Kvikkjokk at the thrust-front in Sweden. Based upon Nicholsen & Rutland (1969) in Norway, and Kulling (1982) in Sweden. Arrows show parts of the section which have been shown to be inverted (see text). The vertical scale is somewhat exaggerated, and the deeper structure is hypothetical.

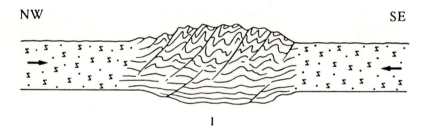

1

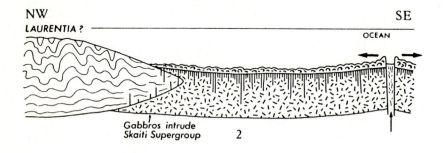

2

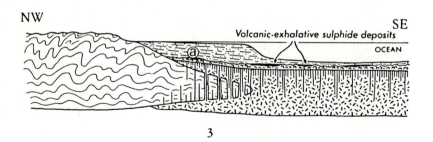

3

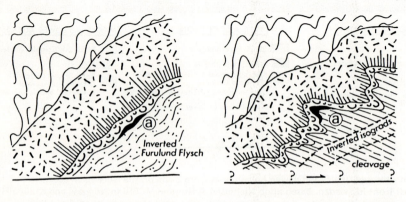

4   5

In the type sections through the Upper Allochthon, in Västerbotten, Sweden, the lower part is dated by fossils and is upright (Kulling 1972, Gee & Zachrisson 1979). The upper parts of the sequence could be upside-down, as at Sulitjelma, implying a synclinal structure for the whole tectono-stratigraphical unit. Binns & Gayer (1980) have proposed a synclinal structure for the Upper Allochthon at Guolasjav'ri, Troms Province, Norway, and Guezou (1978) has identified pre-Scandian metamorphism and deformation in the Gula Group of the Trondheim Nappe near Dombas, Norway.

The upper part of the Upper Allochthon is characterised throughout the Central Scandinavian Caledonides by basic volcanic and intrusive rocks, flysch-like sediments resembling the Furulund Group, and economic deposits of copper and zinc sulphides (Zachrisson 1980). It has generally been regarded as a number of upright thrust-sheets, separated by thrust planes with displacements of at least tens of kilometres. I suggest that the Upper Allochthon may be one thrust-sheet containing major recumbent folds of early Scandian age, the anticlines having pre-Scandian basement in their cores (e.g. the Skaiti Supergroup of Sulitjelma and the Gula Group of Trondheim). The ophiolites represent basic magma which was intruded through this basement during pre-Scandian rifting. Large basic intrusions without associated volcanic members of the ophiolite suite, such as Fongen-Hyllingen (Wilson et al. 1981), may represent intrusions of this magma which failed to break through to the bed of the sea. The rift may have been part of the main Iapetus rift, or a smaller flanking rift.

The recognition of large-scale stratigraphical inversion and pre-Scandian deformation at Sulitjelma has permitted an integrated interpretation of the structure, petrology and economic geology of this classic Caledonian locality. I suggest that it is worth looking for comparable relationships in other parts of the Scandinavian Caledonides.

Fig. 3

Sketch cross-sections through continental and oceanic crust, down to the Moho, to illustrate the evolution of the Skaiti Supergroup, Sulitjelma ophiolite and Furulund Group in the Sulitjelma area.

(1) Deformation and matamorphism of the Skaiti Supergroup in a convergent tectonic setting. Precambrian or early Caledonian (Finnmarkian). The structure shown is extremely schematic.

(2) Initiation and spreading of the rift of the Sulitjelma ophiolite (perhaps Lower Ordovician?). It is thought that the continental margin evolved by thinning, but subsequent Scandian deformation and metamorphism have prevented the recognition of listric faults (Bally et al. 1981) and they are therefore not shown. The continental crust was also extended by intrusion of dykes into the Skaiti Supergroup, accompanied by extrusion of pillow-lavas above the crust, suggesting a similarity with the structure proposed for the Rockall Trough continental margin by Roberts & Montadert (1980).

(3) Later stages in the evolution of the ophiolitic rift. Ore-bodies form above youngest pillow-lavas, gabbros and dykes continue to intrude continental margin (the Skaiti Supergroup), Furulund flysch fills the rift. (a) indicates the part of the section shown on a larger scale in (4) and (5).

(4) Inversion of lower part of Skaiti Supergroup, ophiolite, and Furulund flysch, during the early stages of the evolution of the Sulitjelma overfold. (a) as in (3). It is suggested that the inverted part of the sequence lies above a major thrust horizon separating the Köli and Seve divisions of the Upper Allochthon (M. B. Stephens, *personal communication*, 1983). Early Scandian deformation event.

(5) Imposition of cleavage in Furulund flysch, tight to isoclinal folding of flysch-ophiolite boundary and ore-bodies, inversion of isograds during later stages of evolution of the Sulitjelma overfold. (a) as in (3) and (4). Main Scandian deformation event, Llandovery-Wenlock.

Acknowledgement

I thank A. P. Boyle, T. S. Hansen, W. L. Kirk and R. Nicholson for valuable discussion, the Sulitjelma Mine Company for accommodation and transport in Sulitjelma, and the Central Research Fund, University of London, for financial support. C. F. Stuart drew the figures, and I am particularly grateful for his help with Fig. 3.

References

Bally, A. W., Bernoulli, D., Davis, G. A. and Montadert, L. (1981): Listric normal faults: In Géologie des marges continentales, Proc. Int. Geol. Cong. 26, Paris, C.3., 87–101.
Binns, R. E. (1978): Caledonian nappe correlation and orogenic history in Scandinavia north of latitude 67° N: Geol. Soc. Am. Bull. v.89, 1475–1490.
Binns, R. E. and Gayer, R. A. (1980): Silurian or Upper Ordovician fossils at Guolasjav'ri, Troms, Norway: Nature v.284, 53–55.
Boyle, A. P. (1980): The Sulitjelma amphibolites, Norway: part of a Lower Palaeozoic ophiolite complex?: In Panayiotou, A., ed., Ophiolites: Proc. Int. Ophiolite Symposium, Cyprus 1979: Cyprus Geol. Surv. Dept., Nicosia, 567–575.
Boyle, A. P., Griffiths, A. J. and Mason, R. (1979): Stratigraphical inversion in the Sulitjelma area, Central Scandinavian Caledonides: Geol. Mag., v.116, 393–402.
Boyle, A. P., Hansen, T. S. and Mason, R. (1984): A new tectonic perspective of the Sulitjelma region: In Gee, D. G. & Sturt, B. A., eds., Proc. Uppsala Caledonide Symposium, Wiley, New York, (in press).
Furnes, H., Roberts, D., Thon, A. and Gale, G. H. (1980): Ophiolite fragments in the Scandinavian Caledonides: In Panayiotou, A., ed., Ophiolites: Proc. Int. Ophiolite Symposium, Cyprus 1979: Cyprus Geol. Surv. Dept., Nicosia, 582–600.
Gale, G. H. and Roberts, D. (1974): Trace element geochemistry of Norwegian Lower Palaeozoic basic volcanics and its tectonic implications: Earth Planet. Sci. Lett. v.22, 380–390.
Gayer, R. A. (1973): Caledonian Geology of Arctic Norway: In Pitcher, M. G., ed., Arctic Geology: Am. Assoc. Petr. Geol., Oklahoma.
Gee, D. G. and Zachrisson, E. (1979): The Caledonides in Sweden, Sveriges Geol. Unders., Ser. C 179, 1–48.
Gee, D. G. and Wilson, M. R. (1974): The age of orogenic deformation in the Swedish Caledonides: Am. J. Sci. v.274, 1–9.
Guezou, J. C. (1978): Les tectoniques superposées dans la nappe de Trondheim: Caledonides de Norvège centrale: C.R. Acad. Sci. Paris v.286, Ser.D., 1137–1140.
Henley, K. J. (1970): The structural and metamorphic history of the Sulitjelma region, Norway, with special reference to the nappe hypothesis: Norsk Geol. Tidsskr., v.50, 97–136.
Kirk, W. L. and Mason, R. (1984): Facing of structures in the Furulund Group, Sulitjelma, Norway: Proc. Geol. Assoc. London v.95, 43–50.
Kulling, O. (1972): In Strand, T. and Kulling, O., eds., Scandinavian Caledonides: Wiley, New York.
Kulling, O. (1982): Översikt över södra Norbottensfjällens Kaledonberggrund: Sveriges Geol. Unders., Ser. Ba. v.26, 1–295.
Mason, R. (1971): The chemistry and structure of the Sulitjelma gabbro: Norges Geol. Unders. v.269, 108–141.
Mason, R. (1978): Petrology of the Metamorphic Rocks: Allen & Unwin, London.
Mason, R. (1980): Temperature and pressure estimates in the contact aureole of the Sulitjelma gabbro: implications for an ophiolite origin: In Panayiotou, A., ed., Ophiolites: Proc. Int. Ophiolite Symposium, Cyprus 1979, Cyprus Geol. Surv. Dept., Nicosia, 576–585.
Mason, R. (1981): A trondhjemite vein in the Sulitjelma Gabbro, Norway, and its implications for the age of the Sulitjelma ophiolite: Geol. Mag. v.118, 525–531.
Mason, R. (1984): Inverted isograds at Sulitjelma, Norway: the result of shear-zone deformation: J. Metam. Geol., 2, 77–82.
Nicholson, R. (1974): The Scandinavian Caledonides: In Nairn, A. E. M. and Stehli, F. G., eds., The Ocean Basins and Margins, V.2.: the North Atlanctic, Plenum Press, New York.
Nicholson, R. and Rutland, R. W. R. (1969): A section across the Norwegian Caledonides: Bodo to Sulitjelma: Norges Geol. Unders. v.260, 1–86.

Roberts, D. and Sturt, B.A. (1980): Caledonian deformation in Norway: J. Geol. Soc. London v.137, 241–250.

Roberts, D., Thon, A., Gee, D.G. and Stephens, M.B. (1981): Scandinavian Caledonides: Tectonostratigraphy Map, IGCP Uppsala.

Roberts, D.G. and Montadert, L. (1980): Contrasts in the structure of the passive margins of the Bay of Biscay and Rockall Plateau: Phil. Trans. Roy. Soc. Lond. A. v.294, 97–103.

Turner, F.J. (1981): Metamorphic petrology, McGraw-Hill, New York.

Vogt, T. (1927): Sulitjelmafeltets geologi og petrografi: Norges Geol. Unders. v.121, 1–560.

Wilson, J.R., Esbensen, K.H. and Thy, P. (1981): Igneous petrology of the synorogenic Fongen-Hyllingen layered basic complex, south-central Scandinavian Caledonides: J. Petrol. v.22, 584–627.

Wilson, M.R (1971): The timing of orogenic activity in the Bodo-Sulitjelma tract: Norges Geol. Unders. v.269, 184–190.

Wilson, M.R. (1973): The geological setting of the Sulitjelma ore bodies: Econ. Geol. v.68, 307–316.

Wilson, M.R. (1981): Geochronological results from Sulitjelma, Norway: Terra Cognita v.1., 82.

Zachrisson, E. (1980): Aspects of stratabound base metal mineralisation in the Swedish Caledonides: Geol. Surv. Ireland Spec. Paper No.5, 47–51.

Revised manuscript received 7 Febr. 1984

Caledonide-Appalachian Tectonic Analysis and Evolution of Related Oceans

A. J. Barker[1]/ R. A. Gayer[2]

1 Department of Geology, The University, Southampton SO9 5NH, U.K.
2 Department of Geology, University College, P.O. Box 78, Cardiff CF1 1XL, U.K.

Key Words

Caledonian
Appalachian
Terranes
Laurentia
Baltica
Avalonia
Gondwanaland
Ophiolite obduction
Microcontinent
Collisional tectonics
Iapetus
Rheic Ocean

Abstract

The Caledonide-Appalachian orogen is an example of a complex mountain chain produced by collisional tectonics. The analysis of the evolution of the belt is hindered by ca. 2,000 km of late Caledonian sinistral transcurrence, by superposition of Late Palaeozoic orogenesis, and by post Palaeozoic Atlantic spreading, all of which have disrupted the orogen into geographically and geologically discrete terranes. The geological history of each geographic region is described separately so as to define the individual terranes and to relate them to the evolution of the Iapetus and Rheic oceans. Twenty terranes are identified; five are associated with the SE margin of Iapetus and eight with the NW margin. The remaining seven terranes seem to have been developed within the oceanic domain of Iapetus. Some are microcontinental fragments either rifted from the neighbouring continental margin early in the history of the ocean or separated from one margin, later in the cycle to be drifted across to the opposite margin by subduction of the intervening oceanic lithosphere. Others are clearly remnants of oceanic lithosphere obducted onto adjacent continental margins. The obduction is thought to be a collisional process resulting from the closure of a marginal ocean basin.

The analysis has allowed the construction of eight palaeotectonic maps that record the evolution of the orogen. The rifting and spreading stages of Iapetus occurred from late Riphean through to early Ordovician and at this time the ocean may have been 10,000 km wide in the south between 'Laurentia' and 'Gondwanaland'. A second narrow ocean — Tornquist's sea may have separated 'Baltica' from 'Gondwanaland' along the southern margin of Iapetus. The Finnmarkian orogeny resulted from the collision of a microcontinent with northern 'Baltica' in mid to late Cambrian times, whilst the Grampian and Taconic orogenies were developed when microcontinents collided with 'Laurentia' in latest Cambrian and mid Ordovician times respectively. Iapetus closed as a result of marginally directed subduction systems operating during mid Ordovician through late Silurian times. This resulted in early Silurian continental collision in the Arctic between East Greenland and East Svalbard, in mid Silurian collision between 'Laurentia' and 'Baltica' and in late Silurian/early Devonian collision between 'Laurentia' and 'Avalonia' in the south. Avalonia was separated from 'Gondwanaland' in early Silurian time by spreading of the Rheic ocean, the later history of which spans the interval between Caledonide-Appalachian events and Hercynian-Alleghanian events.

Introduction

The Caledonide-Appalachian orogen extends in a NE-SW direction from northernmost Scandinavia, through Britain and the eastern seaboard of North America to Alabama. In a pre-Atlantic drift configuration the orogen exceeds 7,500 km in length and is normally less than 500 km in width, thus conforming to the general pattern of narrow, elongate intra-continental orogenic belts. The orogen is characterised by a late Precambrian through early Palaeozoic stratigraphy deformed variously in early Palaeozoic time by dominantly NW-SE compression and culminating in closing tectonic events of Devonian age.
In detail the stratigraphy and nature and timing of the deformational events vary considerably both along and across the belt (Fig. 1). The recognition of distinct Cambro-Ordovician faunal provinces within miogeoclinal sequences to either side of the belt led J. Tuzo Wilson (1966) to propose the existence of a former early Palaeozoic ocean, the closure of which gave rise to the fold belt. This ocean was named Iapetus by Harland and Gayer (1972). The presence of ophiolites, representing fragments of oceanic lithosphere (eg. Church and Stevens 1971) and of calc-alkaline volcanic sequences, representing island arc assemblages (eg. Wilkinson and Cann 1974) in distinct terranes separated tectonically from each other and from the bordering miogeoclines, allowed elaboration of the oceanic concept.
Williams and Hatcher (1982) have utilised the concept of suspect terranes (Coney et al. 1980) to construct an accretionary model for the Appalachian part of the orogen. In this approach a suspect terrane is regarded as having been displaced from its original site of formation by transport either perpendicular to the orogen margin as a result of thrusting or subduction of intervening lithosphere, or by strike-slip movements parallel to the length of the orogen. A suspect terrane can be recognised by its sharp structural boundaries with neighbouring terranes and by its distinct geological history up to the time of its accretion.
In the following analysis of the Caledonide Orogen we describe the essential geological features of the terranes to either side of the Iapetus suture. As a result of this analysis certain critical relationships in each of the terranes emerge that constrain models for the evolution of the belt. This model is presented as a series of palaeotectonic maps.

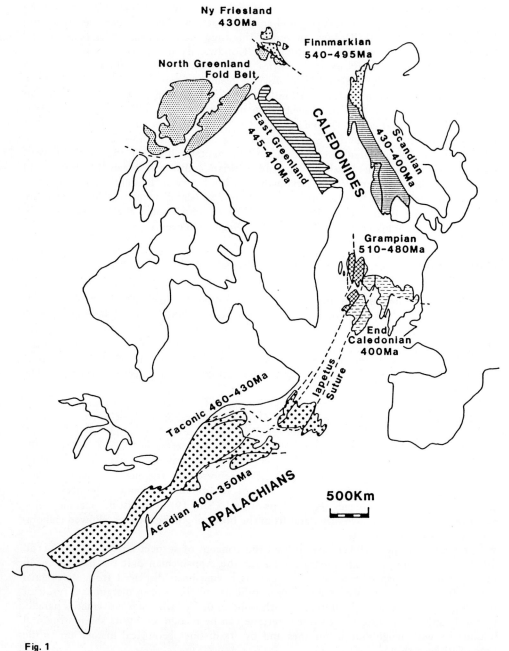

Fig. 1
Regions of the Caledonian-Appalachian Orogen in their pre-Mesozoic drift configurations, showing ages of principal deformation events.

Large dots — Appalachians; dashes — paratectonic British Caledonides; cross-hatch — orthotectonic British Caledonides; closely spaced horizontal lines and triangles — Scandinavian Caledonides; widely spaced horizontal lines — East Greenland Caledonides; small dots — North Green fold belt, Ellesmere Island and West Spitsbergen; open circles — East Spitsbergen.

Regional Descriptions

Areas South and Southeast of the Iapetus Suture

Scandinavia

Rocks deformed in the Caledonian Orogeny occupy a belt 200–400 km wide along the western edge of Scandinavia, extending some 1 800 km from Varangerfjord in the northeast to Stavanger in the southwest. The main deformation has resulted from the collision of the Baltic craton with Laurentia following the closure of Iapetus. With a few notable exceptions, the rocks were developed at the margin of the Baltic craton to the southeast of Iapetus, so that the majority of the rocks developed along the Laurentian margin must now either underlie the Norwegian Sea or be represented to the northwest of the Atlantic.

The Caledonian deformation in Scandinavia produced a series of thrust sheets and nappes which in general were transported southeastwards onto an undeformed cover to the Baltic craton. Hossack and Cooper (in press) have recognised four main allochthonous units which they have correlated along the length of the belt (Fig. 2). These are, from the foreland towards the hinterland, the Baltic Cover Sheets, the Crystalline Sheets, the Oceanic Sheets and a zone of Exotic Nappes. The Baltic Cover Sheets consist of low-medium grade metasediments originally deposited on the sialic crust of the Baltic craton. Slices of this Precambrian basement have been thrust over the cover metasediments and constitute the Crystalline Sheets. The Oceanic Sheets represent rocks originally formed along the southeast margin of Iapetus. Restoration of balanced sections has led Hossack (this volume) to suggest that these Oceanic Sheets may have been transported 400 km from their former sites. The Exotic Nappes are thought to have originated either from microcontinental regions within Iapetus, or from its northwestern margin.

The oldest sedimentary rocks of the Baltic cover occur in Finnmark, N. Norway. Two distinct sequences are developed separated by a major strike-slip fault, the Trollfjord-Komagelv fault (T.-K.F.) The Barents Sea Group, to the north of the T.-K.F., is a Riphean sequence 9 km thick (Siedlecka and Siedlecki 1967), consisting of a lower turbidite sequence passing up into shelf carbonates (Johnson et al. 1978). These are overlain disconformably by the Løkvikfjell Formation consisting of 4 km of fluvial sediments. Both units are intruded by tholeiitic dykes dated at 640 ± 19 Ma (Beckinsale et al. 1976). To the south of the fault is a 2 km thick Riphean sandstone dominated sequence (Siedlecka and Siedlecki 1971; Banks et al. 1974) containing shales which have been dated at 808 ± 19 Ma by Pringle (1973, recalculated to $^{87}Rb\lambda = 1.42$) and capped by a stromatolitic carbonate unit. This is overlain with slight angular unconformity by a younger sequence 3 km thick (Reading 1965) with similar lithologies of Vendian to Tremadoc age. The lower part of the sequence contains the type Varangian tillites which have yielded dates of 654 ± 7 Ma (Pringle 1973, recalculated), and the upper part of the sequence contains a Cambrian-Tremadoc fauna of Baltic affinity. Latest Riphean rocks also occur in the Baltic Cover Sheets of Central Scandinavia and Southern Norway where the Hedmark Group consists of 3 km of fluvial beds, shallow marine sands and turbidites followed by local fissure eruptions of alkali basalt. These are overlain by Varangian tillites and fluvial sands of Vendian age prior to the early Cambrian regional transgression. The Hedmark Group is thought to have developed in an extensional basin possibly related to an early rift phase of the Iapetus Ocean (Nystuen 1982).

At the Caledonian front in southern Finnmark there is only a thin autochthonous sequence present, known as the Dividal Group (Hyolithus Zone). In Finnmark this is of

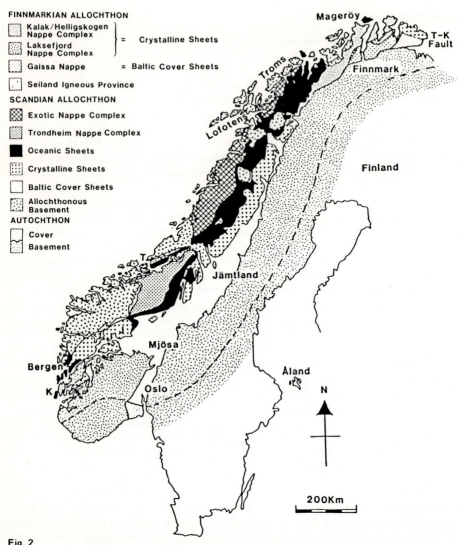

Fig. 2
Simplified tectonic map of the Scandinavian Caledonides, modified from Cooper and Hossack (in press). J = Jotunheim; K = Karmöy; T = Trondheim; T—K = Trollfjord Komagelv. Heavy dashed line = inferred pre-erosion of thrust front.

Cambrian age, and forms a narrow continuous strip along the entire eastern edge of the fold belt, although farther south in central and southern Scandinavia the autochthonous sequence ranges up into Lower Silurian. Lower Cambrian deposits are dominated by sandstones and siltstones, with middle and Upper Cambrian strata being for the most part bituminous black shales, with a varied benthonic and hemipelagic fauna. These shales extend into the Tremadocian, but with generally many breaks in a condensed sequence (Vogt 1967; Føyn 1967). With the exception of the region to the north of the T.-K.F., all the sequences described lie unconformably on the Precambrian Baltic Shield basement. The age of this shield, beneath the Caledonian cover, is usually Mid-Proterozoic (1 500—

1 900 Ma), although in north Norway, Caledonian rocks overstep onto Archaean migmatite with ages of 2 700–2 800 Ma. Similarly in the Lofoten isles ages of 2 800 Ma have been obtained, whilst in south Norway and southwest Sweden, Sveco-Norwegian = Grenvillian ages (930–1 150 Ma) are typical. Unlike most areas on the northwest margin of Iapetus, the Scandinavian area on the southeast side of the ocean did not have a thick carbonate shelf sequence deposited from Cambrian to early Ordovician times. The only extensive area of Cambrian carbonates is the Oslo-Mjøsa region in south Norway, but by latest Cambrian this too had reverted to mudrock deposition. It may suggest that this area was a raised platform in a generally deeper water environment. Certainly the Balto-Scandinavian basin seems to have been a deep water basin, consisting of graptolitic shales. The presence of coarser grained Cambrian sediments on the isle of Åland suggests a coastline in the west Finland area, presumably with appreciable clastic supply.

From mid Cambrian to early Ordovician times the Finnmarkian orogeny affected north Norway, and resulted in the eastward transport of thrust sheets consisting of basement and Riphean to Tremadocian meta-sedimentary rocks broadly comparable to the cover sediments described above. The fact that continental basement slices are found throughout the nappe pile confirms an ensialic nature for the original sedimentary basin. Except for the Magerøy nappe of northwest Finnmark which was transported during the main Scandian orogeny (ca. 420 Ma) (Gee 1975), all other nappes in Finnmark were emplaced during the earlier Finnmarkian orogeny (540–490 Ma) (Sturt et al. 1978). The Finnmarkian Helligskogen nappe extends into north and east Troms, at the base of the nappe pile, but its extension into south Troms is debatable. It seems therefore that the Finnmarkian orogeny *sensu stricto* is restricted to Finnmark and north Troms although a slightly younger Ordovician tectonothermal event seems to have affected parts of central and southern Scandinavia, possibly related to a phase of ophiolite obduction (see below). During the orogeny strong polyphase folding and regional metamorphism occurred. The lower thrust sheets show only low grade metamorphism, but the higher nappes of the Kalak nappe complex show amphibolite facies metamorphism to sillimanite grade, and local development of migmatites. The syn-orogenic Seiland igneous province is the largest area of plutonic rocks associated with the Finnmarkian orogeny, and consists essentially of gabbros and basic dykes but with major ultramafic intrusions and subordinate alkaline intrusions and carbonatites (Robins and Gardner 1975). It is the radiometric ages of 540–490 Ma obtained from rocks of this province which broadly define the timing of the orogeny (Sturt et al. 1978).

At a similar time to the orogeny of north Norway, to the west of present day southwest Norway there was southeasterly directed obduction of ophiolites, such as those of Karmøy and the Bergen Arc, now incorporated in the Oceanic Thrust Sheets of Hossack and Cooper (in press). The Karmøy ophiolite is a particularly good example and shows the full ophiolite stratigraphy from cumulate ultramafics up to hemipelagic metasediments consisting of phyllites and cherts (Sturt et al. 1980). This ophiolite underwent mid to upper greenschist metamorphism prior to uplift and erosion, and before an Upper Ordovician transgressive sequence on south Karmøy. Granites intruding the ophiolite give mid and late Ordovician ages (Furnes et al. 1980), and, since initial Sr ratios indicate a crustal source, it is likely that the ophiolites were obducted onto continental crust prior to mid Ordovician.

Similar, but less well documented, occurrences of ophiolitic obduction occur farther north in Norway, eg. Sulitjelma Ophiolite (Boyle et al. 1979) but the Støren and related ophiolites of the Trondheim area should perhaps be classed separately, suice they underly the Trondheim supergroup which contains a fauna with N. American affinity that suggests these rocks probably originated to the northwest of the Iapetus suture (eg. Gee 1975). During final closure of the ocean the Trondheim rocks were transported southeastwards onto the southeast margin of Iapetus. This ophiolite is unconformably overlain by mid Arenig sediments so that ophiolite obduction, deformation and metamorphism occurred prior to the Arenig. The mid Arenig sediments contain both volcaniclastic and continental detritus, the latter being derived from the northeast (Ryan et al. 1979), which imposes constraints on the palaeogeographic and tectonic setting of the Trondheim supergroup in early Ordovician times.

Even in the stable areas of southwest Norway and south Sweden there are numerous breaks in deposition throughout the Ordovician, many of them probably resulting from transgression and regression related to tectonic movements in the west.

After the orogeny and ophiolite obduction which lasted until approximately Arenig times there was a long period of marine deposition until late Llandovery/early Wenlock times, which marks the age of youngest pre-deformation fossils found in the belt. Lithologies deposited were largely shales, sandstones and limestones, now mostly metamorphosed, but interbanded with these in many areas there are contemporaneous basalts (now amphibolites) and other volcanic rocks. The chemistry of the amphibolites varies from region to region, and even whithin the same thrust sheet. However, most typically they are shown to be either ocean floor tholeiites with a plate margin setting, or calc-alkaline basalts with island arc affinities. This clearly suggests subduction to the west of the Baltic plate at least during some parts of the mid Ordovician to mid Silurian history (eg. Gale and Roberts 1974; Stephens 1979, 1984). It is difficult to obtain tighter controls on the age of igneous activity in many areas of the belt, especially within the Oceanic Thrust Sheets. This is because regional metamorphism has reset the isotopic systems and obliterated most fossils that may have been present. Faunal provinciality was greatest in the early Ordovician, but from this time onwards it slowly became less pronounced until by latest Ordovician there was much greater interchange of fauna (Bruton and Lindstrom 1981). This is thought to reflect a narrowing/shallowing ocean. By Wenlock times marine sedimentation had ceased, being latest in the east and southeast, in front of the advancing Caledonian orogeny. In Jämtland a Llandovery marine sequence is followed by an early Wenlock, Old Red Sandstone molasse facies (Bassett et al. 1982). This progrades over other areas to the south and east during Wenlock and Ludlow times. The Ordovician to Silurian autochthonous sequences of Sweden were deposited in the generally shallow Balto-Scandian basin, and are very different from the deeper-water facies to the west. They are almost completely lacking in contemporaneous volcanic rocks and are dominated by shallow water carbonates, in contrast to the earlier clastic sequences of the Cambrian (eg. Larsson 1973).

The Scandian phase of the Caledonian orogeny began in mid Silurian times, continuing to late Silurian. It produced regional metamorphism to sillimanite grade, and local migmatization. In southwest and west Norway eclogites are found which have been shown to form a progressively higher pressure and temperature westwards (ie. towards the suture) (Bryhni and Andreasson 1981). These suggest westward subduction of the Baltic plate to depths of at least 50 km.

A polyphase deformation and metamorphism occurred throughout the belt, and involved several phases of intense early isoclinal folding. The eastward directed thrusting probably began during the early phases of deformation but the final movements generally post-date the main schistosity and are therefore fairly late. Erosion of major late, open, upright folds that are superimposed on the earlier structures, has produced extensive windows in which the lower tectonic units are exposed and allow over 400 km of eastward nappe

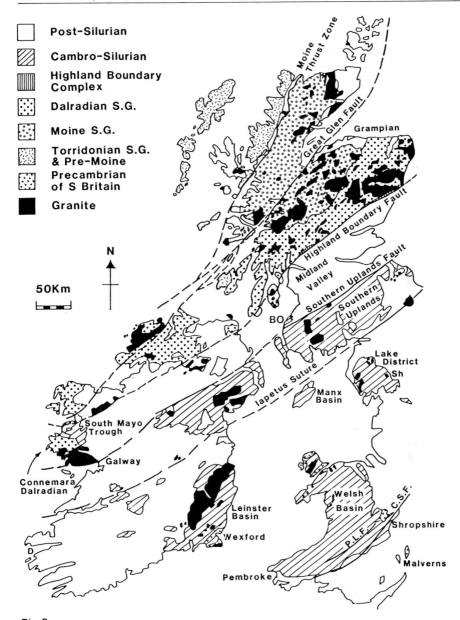

Fig. 3
Simplified map of the British Caledonides, modified from Leake et al. (1983).
D = Dingle Bay; Sh = Shap granite; P.L.F. = Pontesford-Linley Fault; C.S.F. = Church Stretton Fault.

translation to be deduced (eg. Gee 1975). The syn-sedimentary basalts (now amphibolites) within the Scandinavian allochthon have been discussed briefly above, but there were also many post-sedimentation plutonic rocks that show pre- or syn-orogenic relations. These include gabbros, diorites, trondhjemites, granites and ultrabasic rocks, and show a variety of ages. Of these the granitoid rocks are the most abundant. Trondhjemites and granites give a range of ages from mid-Ordovician to late Silurian, and in the south central region several phases have been recognised. The earliest of these were probably generated at the time of the Finnmarkian orogeny.

The orogeny in Scandinavia was virtually completed by the earliest Devonian, and from this time until the mid Devonian, in coastal areas of Norway several variably sized intermontane fault bounded extensional basins were developed. These basins, filled with Old Red Sandstone facies molasse deposits, were later disrupted by mid Devonian thrusting and folding, which has been related by Roberts (1983) to the major sinistral transcurrent fault system that extended from Svalbard to the eastern Appalachians.

Britain South of the Solway Firth — Shannon Suture

In south Britain, resting on Precambrian basement which consists of schists, gneisses and plutonic rocks, are isolated Precambrian sequences consisting essentially of interbedded clastic and volcanic rocks (eg. Wexford, Pembroke, Shropshire, Malverns and Charnwood) (Fig. 3). The basement of southeast Ireland is probably older than 1600 Ma (Max 1975) whilst that in the Anglo-Welsh area appears to be no older than late Proterozoic (1200 Ma) (Hampton and Taylor 1983). The overlying sediments, intrusive and extrusive rocks of different areas consistently give late Precambrian ages in the 600–700 Ma range (eg. Fitch et al. 1969, Meneisy and Miller 1963, Lambert and Rex 1966). Lying unconformably above the Precambrian rocks there is a marine Cambrian sequence which consists of shallow water clastic deposits bordering the Midland Platform to the southeast, as well as a mixture of deep and shallow water sediments in unstable shelf areas such as the subsiding Welsh Basin. This sequence consists essentially of shales and sandstones, some of which are turbidites, and contains a fauna largely of trilobites and brachiopods, of the Gondwanaland province (see Dean, this volume). These also show similarities with faunas of the Baltic area, indicating that these areas lay to the southeast of the Iapetus Ocean.

During early Ordovician times there was a major Arenig transgression, probably comparable to that of similar age seen in southern Scandinavia. Also during the Tremadoc-Arenig, major volcanism commenced in the Welsh area, with the height of activity being in the Llandeillo and Caradoc of north Wales. Major activity also occurred in the Lake District during the Llandeillo, but by early Caradoc times the main activity was concentrated in north Wales and the southern part of the Leinster Basin of Eire. Fitton and Hughes (1970) demonstrated a trend from tholeiitic to calc-alkaline volcanism southeastwards across the Lake District, and thus suggested a southeasterly dipping subduction zone beneath this area. Stillman et al. (1974) have documented a similar situation in southeast Eire.

Sedimentation in the basinal areas throughout the Ordovician was essentially of deep water graptolitic shales with similar amounts of sandstones and volcanic rocks. It is important to note that the southern Britain area did not experience the deformation, metamorphism and plutonism that occurred in north Norway during the mid Cambrian to early Ordovician. Neither does it show evidence for early Ordovician ophiolite obduction as is observed in central and southern Norway. In contrast, the Welsh area, during the early-mid Ordovician, was the site of a fault-bounded extensional basin in which crustal thinning was sufficient to allow major volcanism. This basin was probably developed as a marginal basin situated behind an island arc system. It is probable that movement along some of the extensional faults initiated turbidite flows and slumps. Work by Simpson (1968) has shown that late Llanvirn to Caradoc folding/warping affected rocks of the Lake District and Manx basins. It is clear therefore that, although southern Britain was not affected by the strong orogenic affects that north Norway experienced during mid Cambrian to early Ordovician, it was by no means a stable area during the early and mid Ordovician.

The late Ordovician evolution of southern Britain involved the deposition of more graptolitic shales, followed by the deposition of thick sequences of Silurian turbidites in the basinal areas (eg. Aberystwyth Grits of the Welsh basin). These have a fauna of brachiopods and graptolites, and in the nearshore shelf areas a rich assemblage of brachiopods, crinoids and corals are found. Sediments of the Welsh and Lake District basins range up to the Whitcliffian stage of the Ludlow series, and in the Lake District southerly directed palaeocurrents have been demonstrated. This clearly indicates the presence of a nearby northern landmass by this time, although there is no evidence that this was the Southern Uplands area that is at present to the north; transcurrent movements may have subsequently juxtaposed them. Certainly the ocean had narrowed considerably from early Ordovician times when faunal provinciality was at its greatest, and by early Silurian times there was considerable mixing of fauna from both sides of Iapetus.

Waning volcanicity in southern Britain continued throughout the late Ordovician and early Silurian. The termination of volcanism is progressively later to the west, with the youngest volcanicity being of Wenlock age in the Dingle area of W. Eire. Because of this trend, Phillips et al. (1976) proposed that there was a southwesterly migrating point of collision resulting from oblique plate vectors.

Deformation and metamorphism occurred in southern Britain from mid to late Silurian, and major upright folds were established in the basinal areas. These are generally oriented NE-SW and have an associated axial planar cleavage. In north Wales, Coward and Siddans (1979), following Shackleton (1954) have demonstrated the presence of a major decollement with 43 km of NW-SE shortening with which these folds and cleavage are associated. Strike slip faults with a similar orientation to that of the fold axes developed at a slightly later stage in the orogeny, followed by local thrusting and development of kink bands. The metamorphism of this end Silurian event is sub-greenschist facies, thus indicating low temperatures. There appears to be no syn-tectonic plutonism, but in latest Silurian and early Devonian times granite plutons with contact metamorphic aureoles were emplaced in Eire and the Lake District (eg. Shap granite).

The relatively gentle end Silurian deformation and metamorphism suggests that southern Britain was probably situated in a region where continental convergence, following closure of Iapetus, was not sufficient to generate Himalayan style orogenesis as appears to be the case in Scandinavia.

Newfoundland to the Appalachians

The N. American section of the Caledonian-Appalachian orogen has a present day NE-SW trend, and within this belt a number of structurally and stratigraphically distinct regions or terranes with similar NE-SW trends can be recognised (Williams and Hatcher 1982). For simplicity three main regions can be identified (Fig. 4). The first region extends along the northwest margin of the belt and consists of the Appalachian miogeocline,

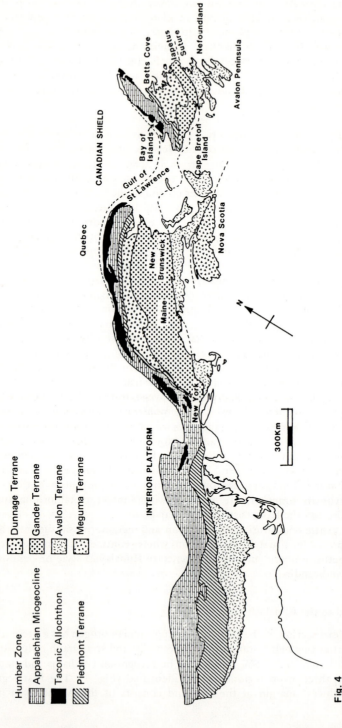

Fig. 4
Simplified map of the Appalachians and Newfoundland, showing the main tectonic terranes, modified from Hatcher and Williams (1982).

isolated areas of overthrust taconic allochthon, and the Piedmont terrane. This region is often referred to as the Humber Zone (eg. Schenk 1978). The central portion of the belt consists of the Dunnage and Gander terranes, and in the eastern part of the belt is the Avalon terrane. A major fault separates the Avalon terrane from other areas to the northwest. In Newfoundland this is the Dover-Hermitage fault, and represents the line of the Iapetus suture separating regions originally on the southeast side of Iapetus (ie. Avalon) from those on the northwest side. Further southeast from the Avalon terrane is the Meguma terrain of Nova Scotia, but for the purpose of this review it will not be discussed in detail.

1 The Humber Zone (Miogeocline, Taconic allochthon and Piedmont terrane)

The basement to this region is Grenvillian Canadian Shield rocks which experienced a major tectono-thermal event approximately 1000 Ma ago (Schenk 1978), and are cut by northwest trending dolerite dykes. Inliers of basement are seen throughout the Humber Zone as well as forming a major shield area to the northwest. The crystalline basement is overlain by late Precambrian to early Palaeozoic (Cambro-Ordovician) rocks consisting of clastic sediments overlain by carbonates with trilobites of Laurentian affinity. As Williams (1979) suggested, this is probably an extension of the northwest Scotland area of Lewisian basement overlain by Torridonian sandstones, in turn overlain by Cambro-Ordovician carbonates. The Moine and Dalradian of Scotland and Ireland show comparable ages, lithologies and tectono-metamorphic features to the Fleur de Lys Supergroup of Newfoundland (Church 1969, Williams et al. 1974). As with the Dalradian, the Fleur de Lys Supergroup of pelites and psammites is considered to have developed on the continental rise to the southeast of a carbonate bank. This supergroup was highly deformed and metamorphosed in early Arenig times and subsequently thrust northwestward onto the miogeocline platform carbonates. It is generally overlain by a variety of allochthonous slices, including volcanics, tectonic mélange and ophiolites. The ophiolite suites form the uppermost allochthon, with the best known example being the Bay of Islands ophiolite (Fig. 4). The timing of main deformation, metamorphism (which reached amphibolite facies) and ophiolite emplacement is generally considered to be of early mid-Ordovician age (Taconic), since mid Ordovician reefal carbonates unconformably overlie the allochthonous slices in northwest Newfoundland (Bergstrom et al. 1974). In addition ^{40}Ar-^{39}Ar dates indicate 460–480 Ma for the thrusting (Dalmeyer and Williams 1975). However, final metamorphism and deformation probably occurred during the Devonian (Bursnall and De Wit 1975). Associated with the orogenic events there was also minor intrusion of small granitic to ultrabasic plutons, but these are only of minor significance compared with the major (mostly Devonian) granitic plutons of more easterly regions. The eastern margin of the Humber Zone of Newfoundland consists of a thin strip of the Piedmont terrain. These rocks are considered to be deformed and metamorphosed equivalents of rocks deposited at the eastern edge of the Miogeocline. The eastern margin of this terrane in Newfoundland is defined by the Baie Verte lineament, which is marked by deformed ophiolites and basic volcanic rocks. The lineament continues to the Brompton line of Quebec, and Williams and King (1977) interpret this important suture as a relict continental margin-ocean interface.

2 The Dunnage-Gander Terranes

The Dunnage terrane is found intermittently along the length of the Appalachians, but is well represented in central Newfoundland. It consists of Cambrian to mid Ordovician basic volcanic rocks, sediments (cherts, slates, greywackes and minor carbonates) and mélanges that overlie an ophiolitic base (Williams and Max 1979, Hatcher and Williams 1982). It is noticeable that Precambrian strata are absent.

The Dunnage terrane is characteristically weakly deformed and metamorphosed compared to the adjacent Piedmont and Gander terrains (Hatcher and Williams 1982). It is generally considered that the Dunnage rocks represent remnants of oceanic crust and island arcs, and that the mélanges within the terrane are subduction related (e.g. Schenk 1978). Well developed ophiolite suites are present in the west of the zone, such as Betts Cove, Newfoundland (Fig. 4). As noted above, the boundary with the Piedmont terrane is the Baie Verte lineament. However, the exact divide is difficult to determine since ophiolitic complexes of the Dunnage terrane grade structurally and metamorphically into ophiolite mélanges and metasediments of the Piedmont terrane (Hatcher and Williams 1982). As with the Humber Zone, the timing of main deformation in the Dunnage terrane is considered to be mid Ordovician. The orogeny had clearly terminated by the late Ordovician, since middle to upper Ordovician and Silurian strata overlap all rocks from the miogeocline of Quebec to the Gander terrane of New Brunswick (Williams 1978). This clearly indicates that by this time the Gander and Dunnage terranes were closely connected with each other and the Laurentian continent. Previously, oceanic lithosphere (now partly preserved as Dunnage terrane ophiolites and mélanges) had separated the Gander terrane from the Laurentian continent. The overlapping Upper Ordovician and Silurian sediments exhibit a penetrative foliation produced during the mid-Devonian (Acadian) orogeny.

The boundary between Dunnage and Gander terranes is largely covered by Silurian Strata, but in Newfoundland is represented by sporadic ophiolites and mélanges. The diagnostic feature of the Gander terrane is the presence of continental basement. The southeastern boundary of the zone in Newfoundland is marked by the Dover-Hermitage Fault.

Above the continental basement (migmatites and sheared gneissic granites), the Gander terrane in Newfoundland consists mostly of late Precambrian turbidites. These are tectonised and intruded by early Cambrian (570 Ma) granites (Kennedy 1976). However, Bell et al (1977) have established that many of the granites of this terrane are essentially Devonian (ages of 420–355 Ma). Plutons of this age are not restricted to the Gander terrane, but are also abundant in the Dunnage terrane, as well as being present within the Avalon and Meguma terranes.

In New England, the Gander terrane is characterised by a deformed interbedded pre-mid Ordovician sequence of shales and volcanic rocks above gneissic domes (Robinson and Hall 1980). In the central New Brunswick area, the Miramicki terrane is considered by most to be a Gander terrane correlative. It consists of deformed and metamorphosed Ordovician sediments and volcanic rocks above quartz arenites with an Arenig fauna (Neuman 1972), lying over schists and gneisses. The succession has experienced granite intrusion typical of the Gander terrane but it should be added that no Precambrian continental basement has definitely been identified.

The first deformation within the Gander terrane is considered to be pre-Silurian, and in certain areas has been suggested as pre-mid Ordovician (e.g. Kennedy 1976). However, the peak of metamorphism and deformation is mostly considered to be mid Palaeozoic, associated with the Acadian event (Blackwood 1978; Robinson and Hall 1980).

3 The Avalon Terrane

This terrane consists mostly of Upper Precambrian sediments (including a tillite formation), and volcanic rocks, which show lesser amounts of deformation and metamorphism compared to rocks of adjacent terranes. In some areas Precambrian granitic intrusions cut the sequence. Locally, the Precambrian sequence is observed to pass conformably upwards into a shaly Cambrian sequence. Elsewhere (Newfoundland in particular), there is an angular unconformity, representing a short period of orogeny prior to the deposition of late Precambrian molasse, followed by Cambrian to Lower Ordovician shallow water sediments, mostly shales (Schenk 1978), including an assemblage of trilobites with Gondwanaland affinity. In Newfoundland, the basement to the Avalon terrane is unknown, but it may be present elsewhere in the Maritime Provinces (Kennedy 1976). Sequences of southern New Brunswick and Cape Breton Island show close resemblance to the Avalon stratigraphy of Newfoundland (Schenk 1978), and consist of marble, gneiss and schists overlain by volcanic rocks.

The Gander and Avalon terranes are linked by a common Devonian history of plutonism (Williams and Hatcher 1982). This is associated with the Acadian orogeny which affects all terranes of the Appalachian belt. Locally, Silurian-Devonian sedimentary links can be demonstrated between the Gander and Avalon terranes. Prior to this however, the Avalon terrane shows a very contrasting history to the Gander and more northwesterly terranes. The most important features are the lack of an Ordovician orogenic event, and the contrasting nature of Cambrian trilobite faunas. It is clear therefore that the Avalon area must have been very distant at this time, and most if not all workers would place it on the southeastern side of Iapetus, even if not making up part of the major southeastern continent.

The Union Complex of Maine has been interpreted as ophiolitic, and coupled with the isotopic age of 410 Ma (Gaudette 1981) the Gander-Avalon terrane junction of south Maine has been suggested as a mid-Palaeozoic oceanic suture.

Steep mylonite zones and major faults generally mark the western boundary of the Avalon Zone, suggesting major transcurrent movement to juxtapose the terranes in their present position (Williams and Hatcher 1982). Palaeomagnetic studies by Kent and Opdyke (1980) give strong evidence for major sinistral displacements (circa 1000–2000 km) along this suture after the latest Devonian. Since Bell et al. (1977) have dated a pluton which straddles the fault as 345 Ma, the fault movement can be no later than early Carboniferous, and presumably roughly coincides with the main movement along the Great Glen Fault in Britain.

Areas north and northwest of the Iapetus Suture

Britain north of the Solway Firth – Shannon Suture

For simplicity north Britain can be divided into two regions; an area to the north of the Highland Boundary Fault (H.B.F.) and its extention into Ireland, which experienced major deformation and metamorphism in the early Ordovician, the Grampian orogeny, and referred to as the orthotectonic Caledonides; and an area to the south of this fault which did not (Fig. 3). The area south of the H.B.F. includes the Midland Valley of Scotland and the Southern Uplands, the divide between the two areas being the Southern Uplands Fault (S.U.F.). This also marks an important boundary, but since the histories of these two areas are very much inter-related, they will be considered together as the northern part of the paratectonic Caledonides. The Connemara-Galway area of Eire is a highly metamorphosed and tectonised area of Grampian (Dalradian) rocks in an anomalous position south of the H.B.F. extension. It shows every similarity with the Grampian rocks to the north of the H.B.F. extension but is separated from this area by the Ordovician-Silurian South Mayo Trough, which shows only minor deformation and metamorphism. A possible explanation of this problematical situation is given in the final model and will not be discussed at this stage.

1 The region north of the Highland Boundary Fault

In the foreland area of the northwest Highlands of Scotland 'Torridonian' clastic sediments lie unconformably above Lewisian (Archaean) Basement. The older Stoer Group of the 'Torridonian' is dated at 1100 Ma based on palaeomagnetic data (Smith et al. 1983), and is unconformably overlain by the 7 km thick Torridon Group of red sandstones with marine intercalations deposited in a subsiding fault controlled basin. The lower part of the Torridon Group has been dated at 1040 Ma by Smith et al (1983), and it is suggested by Stewart (1975) that the Torridon basinal sequence may have been associated with the opening of Iapetus. Similar views have also been expressed by Anderton (1982). Lying unconformably on both Lewisian and Torridonian rocks is a Cambro-Ordovician shelf sequence. This commences with a lower Cambrian sandy facies (basal quartzite and pipe rock), and from mid Cambrian to Arenig times is dominated by shelf carbonates (Durness Lst.). The Cambro-Ordovician sequence contains trilobites with Laurentian affinity, and the sequence in general can be directly correlated with that of northwestern Newfoundland and east Greenland. The highly deformed and metamorphosed area of north Britain (orthotectonic Caledonides) lying between the Moine thrust and H.B.F. is divided into two sequences; an older Moine succession and a younger sequence known as the Dalradian Supergroup, into which the Moine appears to pass upwards with conformity.

The Moine consists of a thick monotonous sequence of schists and psammites. Its age is certainly older than 654 ± 7 Ma, which is the age of the Varanger Tillite of north Norway (Pringle 1973) with which the Port Askaig tillite in the mid Dalradian is correlated. Much of the Moine is probably early Riphean, since Brewer et al. (1979) have demonstrated that the earliest metamorphic event of the Moine pelites in the southwest of the Northern Highlands is 1004 ± 28 Ma, correlatable with the Grenvillian orogeny of North America. Brook et al. (1976) have suggested an age of sedimentation for the Moine s.s. of 1250–1050 Ma, somewhat older than the earliest Torridonian rocks of northwest Scotland, with which they have been traditionally correlated. It has been demonstrated by Powell (1974) and by more recent workers that the "older moines" underwent a tectono-thermal event c. 730 Ma ago (the "Morarian") although more recent data casts doubt on the tectonic significance of this event (Powell et al. 1983). Originally this event was thought to be confined to the Moines north of the Great Glen Fault (G.G.F.), but Piasecki and Van Breeman (1979) and more recent work by Piasecki has shown that the Central Highland Division south of the G.G.F. rests on a Grenvillian basement and also exhibits a Morarian event. The Central Highland Moines pass up into the Dalradian Supergroup, estimated by Harris and Pitcher (1975) to have a thickness of about 20 km. The lowest division (Appin Group) commences with a sequence of interbedded quartzites and pelites with some limestone. Above this is the mid Dalradian (Argyll Group) which contains quartzites and pelites as well as the Port Askaig tillite discussed previously. The tillite is a widespread deposit seen in both Scotland and Ireland, and is likely to correlate with the Varangian tillites of north Norway, Greenland,

Spitsbergen and Newfoundland. It would thus seem to indicate a widespread glacial event in Vendian times, constituting an important chronostratigraphic unit in Caledonian correlation. The upper part of the mid Dalradian (Southern Highlands Group) is of early Cambrian age and consists of an association of limestones and basic volcanic rocks. In the Upper Dalradian there are sandstones, shales and volcanic rocks of the Highland Boundary Complex, including the fossiliferous McDuff slates which yield an early Ordovician, probably Arenig, age (Downie et al. 1971) and the Dounans Limestone which contains a Laurentian, Canadian fauna (= early Arenig) (Curry et al. 1982). There are complications with the correlation of these rocks with the remainder of the Dalradian Supergroup since they are associated with the H.B.F. zone. They may represent a distinct tectonic unit, unrelated to the Grampian Orogeny.

The correlation of the lower part of the sequence is uncertain, but it is believed by many that the mid and upper Dalradian is a deeper water shelf turbidite sequence further offshore from the northwest Scotland shallow water shelf facies.

In early to mid Ordovician times a major orogeny (the Grampian orogeny) affected all rocks north of the H.B.F. The main features of this orogeny are as follows:

The orogeny was initiated after the deposition of the uppermost Dalradian of Scotland, possibly post-early Arenig (but see above). This is younger than the Finnmarkian orogeny of north Norway which began in the mid Cambrian. However, the syn-tectonic Cashel gabbro of Connemara dated as 510 Ma (Pidgeon 1969) suggests an earlier development. Similarly the Ben Vurich granite in Scotland was intruded after D3 of the Grampian deformation but yields an age of 514 ± 7 Ma and suggests latest Cambrian for the initiation of the orogeny. Lower Silurian (Llandovery) deposits in western Eire overlie the deformed and metamorphosed Dalradian, and therefore show that the Grampian event was confined to the Ordovician. The youngest syn-tectonic intrusion of the orthotectonic Caledonides is dated as 480 Ma, and Pankhurst (1974), studying post-tectonic granites of N.E. Scotland, obtained a Rb-Sr whole rock isochron of 460 Ma. This clearly suggests that the Grampian orogeny was completed by mid-Ordovician times.

The orogeny consisted of intense polyphase deformation, regional metamorphism to sillimanite grade, and with migmatization and extensive plutonism in certain areas. The plutons were mostly granitic, but some of the early examples in particular were basic. Deformation included major northwesterly directed thrusts such as those of the northwest Highlands, as well as structures with southeastward verging fold nappes (such as the Tay Nappe) in the southern part of the belt. Most fold axes trend NE-SW, perpendicular to the direction of compression, but the pattern has been greatly complicated by many later phases. The pattern of metamorphism is complex since the Grampian metamorphism oveprints the earlier Morarian features, such that there is complex interference. Grampian metamorphism also consists of more than one phase, and in the Buchan area thermal overprinting further complicates the picture.

Pre-, syn- and post-tectonic plutonism occurred associated with the Grampian event, the later examples all being acidic. The Connemara/Galway granites show initial $^{87}Sr/^{86}Sr$ ratios of 0.706 to 0.710, whilst the granites of Scotland have considerably higher values in the 0.714 to 0.719 range (Brown 1979). These Scottish granites with relatively high initial strontium ratios are unlike granites emplaced at modern destructive margins such as the Andes. To explain these high values, Brown (1979) argues that a high amount of crustal fusion is necessary. This may have resulted from above average crustal temperatures caused by radio-active decay from the pile of Dalradian nappes (e.g. Richardson and Powell 1976).

The cause of the Grampian orogeny is controversial and open to many interpretations. The present authors' consider it to be the result of collision with a micro-continent (the Midland Valley terrane) in the same way that Mattauer et al. (1983) have shown that the North American Cordillera with associated deformation and metamorphism resulted from Jurassic collision of the North American continent with the Stikine terrane. This analogy demonstrates that amphibolite facies metamorphism and polyphase deformation are not necessarily the result of collision between major continents.

The area of Britain north of the H.B.F. contains no mid Ordovician to mid Silurian deposits, and exhibits only extremely mild mid to end Silurian orogenic effects compared to the marked compression in the basinal areas of southern Britain and the intense deformation and metamorphism in Scandinavia. However, late Silurian to early Devonian intrusion of many granitoid bodies occurred in the Irish and Scottish area north of the H.B.F. These have initial $^{87}Sr/^{86}Sr$ ratios < 0.709 (Halliday et al. 1979) and may well be associated with a destructive plate margin (Thirlwall 1981). This margin is considered to exist to the southeast of the Southern Uplands with a northwesterly directed subduction zone, and will be discussed later.

From ? late Silurian to mid Devonian time there was sporadic deposition of Old Red Sandstone facies molasse type deposits and volcanic rocks in variably sized inter-montane basins. Final adjustments related to the Caledonian-Appalachian orogeny gave rise to late Devonian strike-slip movements along the G.G.F. to bring the areas either side of this fault approximately to their present positions. The direction and amount of displacement along the Great Glen Fault has always been a point of contention, but the general concensus is of several hundred kilometres (or more) of sinistral displacement (eg. Watson and Smith 1983) during the mid to late Devonian, followed by a much smaller component of dextral movement in post-Devonian times.

2 The Midland Valley and Southern Uplands

The pre-Arenigian history of these two important areas is uncertain, and open to various interpretations since no rocks older than the Arenig epoch are exposed at the surface. However, Upton et al. (1976 and 1983) have reported xenoliths of garnet gneiss in carboniferous volcanic necks near the northern boundary of the Southern Uplands, thus suggesting a granulite basement, presumably of Precambrian age. Such a basement is thought to extend southwards under the Southern Uplands for at least 15–20 km from seismological evidence (Hall et al. 1983) and possibly to the Solway Suture from the evidence of the 'WINCH' deep seismic reflection profile (Brewer et al. 1983). It seems likely that the Midland Valley represents part of a microcontinent that accreted to the major continent on the northwest margin of Iapetus giving rise to the Grampian orogeny.

The Ballantrae ophiolite which contains sediments yielding an Arenig fauna, is situated just north of the Southern Uplands Fault (Fig. 3). It is considered to be a fragment of oceanic lithosphere obducted to its present position in the late Arenig (possibly early Llanvirn) during the waning stages of the Grampian orogeny. To the south of the Southern Uplands Fault is the Southern Uplands terrane, with a sequence of rocks ranging in age from Llanvirn to Llandovery. The older, Llanvirn and Llandeilo rocks consist of basalts overlain by cherts, and suggest an ocean floor origin. The younger rocks consist essentially of interbedded shales, greywackes, conglomerates and pyroclastic rocks. Since the work of McKerrow et al. (1977) it has generally become accepted that these deposits represent an accretionary prism above a northwesterly dipping subduction zone at the northwestern margin of Iapetus. This model is based on faunal and structural evidence, and it has been demonstrated that a series of imbricates which grouped together show southeastward younging, individually young northwestwards. The accretionary prism developed throughout Ordovician and early Silurian times, and as a result, by Llandovery times the ocean margin had extended 70 km further south from its initial position. Also by Llandovery times, the northwestern margin of the accretionary wedge had emerged above sea-level and gave rise to red bed deposition. This landmass was termed "Cockburnland" by Walton (1963), and resulting from its emergence, molasse deposits spread northwards through the Midland Valley graben, which was also receiving sediment supply from the mountains to the north. The initiation of the Midland Valley graben has not been precisely dated but the concensus is for a post-Arenig, Ordovician age (see discussion by Leake 1978). With the mid to end Silurian closure of Iapetus there was compression across the Southern Uplands area with thrust emplacement of the accretionary prism northwestwards onto the continental margin (Hall et al. 1983). However, it should be emphasised that most deformation in this region was caused during the development of the accretionary prism. There was very little metamorphism associated with this closure, and only locally is greenschist facies metamorphism developed in the Southern Uplands. Late granite plutons occur in the Southern Uplands, but none is found in the Midland Valley.

A sequence of lower Old Red Sandstone facies rocks up to 9 km thick is seen in the Midland Valley. These have been dated by their fish faunas which Westoll (1945) considered to be essentially Devonian and no older than Downtonian (Pridoli). However, Lamont's (1952) reappraisal of the faunas, suggested the possibility of a Llandoverian age. Thus the interbedded volcanics of the Midland Valley could have ages anywhere in the range from earliest Silurian to Devonian and could pre-date the final closure of Iapetus. The studies by Thirlwall (1981) demonstrate SE-NW variation in the chemistry of volcanics that would indicate an active northwesterly dipping subduction zone. If the rocks are of Devonian age this is difficult to explain, but if they are of Llandoverian age it may be explained by activity associated with the closure of Iapetus by subduction beneath the Southern Uplands. However, the lack of plutons in the Midland Valley is a problem yet to be resolved.

Greenland

1 East and Northeast Greenland

Archaean and early Proterozoic basement of Greenland is overlain in central East Greenland by the Krummedal supracrustal sequence, which underwent an orogenic event in the period 1200–900 Ma. In Northeast Greenland, sediments of perhaps comparable age have experienced the Carolinidian orogeny dated at approximately 1000 Ma (Henriksen 1978).

Major developments of late Proterozoic through Vendian to Cambro-Ordovician sediments occur in central East Greenland (Haller 1970). The Eleonore Bay Group includes 13 km of the 17 km sequence, and extends to the base of the Vendian. It is a dominantly arenaceous-argillaceous sequence although in the upper parts carbonates containing Riphean stromatolites become abundant. Above this is 1 km of Vendian rocks composed of tillite, shale and sandstone. These pass into a Cambro-Ordovician sequence commencing with a basal sandy facies, but followed by a 3 km sequence of shelf carbonates. These range in age from early Cambrian to mid Ordovician, although rocks of late Cambrian age appear to be absent and hence suggest a break in sedimentation.

In Northeast Greenland a sequence of sediments (Hagen Fjord Group) extends from latest Riphean to ? early Cambrian age. It is a 5—6 km thick sequence of essentially arenaceous and argillaceous rocks, with minor amounts of carbonates mostly at the top of the sequence. The Cambro-Ordovician carbonate shelf sequence is not present in Northeast Greenland.

Absent from Northeast Greenland, but present in the central East Greenland Caledonides, Devonian red beds lie unconformably above the Cambro-Ordovidian carbonates. Prior to their deposition however, there was an orogenic event which gave rise to deformation, metamorphism and plutonism. The deformation resulted in folding, and westerly directed thrusts. In many areas this folding was superimposed on intense Proterozoic folding of the Carolinidian event, such that only the later more simple structures can be unambiguously assigned to the Caledonian.

The exact timing of the Caledonian tectono-metamorphic event is controversial, and the various lines of evidence put forward are inconclusive. Stratigraphically the event must have occurred after the deposition of the youngest rocks in the shelf sequence (mid-Ordovician) and prior to the deposition of the Devonian red beds. The interpretation of the radiometric data is uncertain. Metamorphism reached amphibolite facies in the so-called "crystalline region" of the southern sector of the East Greenland Caledonides, but elsewhere metamorphism is usually of considerably lower grade. Rb-Sr and K-Ar dates range from 445—410 Ma in the south to 420—375 Ma further north (Higgins and Phillips 1979) and thus give a range of ages from late Ordovician to mid Devonian. This suggests either a very long metamorphic event, or that some or all of the dates may not reflect the age of metamorphic recrystallisation. Only locally was the metamorphism sufficiently intense to upset the earlier Rb-Sr isotope system, so that many of the dates are based on K-Ar determinations and may therefore be underestimates, representing cooling ages. However, this cannot explain the total range of dates, leaving the exact timing of the Caledonian tectono-metamorphic event of East Greenland uncertain.

Caledonian plutonic rocks are largely confined to the southern part of the belt. The earliest are sheets of monzonite and granodiorite emplaced into Proterozoic migmatites at Scoresby Sund. Isochron ages of around 475 Ma have been obtained from these, suggesting an early mid-Ordovician age, whilst a nearby granite gives a zircon age of 510 Ma. Late granites emplaced at the margins of the crystalline complexes give late Ordovician ages of 440—455 Ma. From this limited data Higgins and Phillips (1979) suggested that early granites (c. 475 Ma) intruded prior to the end of sedimentation, and perhaps that peak metamorphism was shortly after. The late granites (c. 450 Ma) show syn- and post-folding relationships and would indicate tectonism in the mid to late Ordovician. Thrust movements and abundant faults appear to have occurred at a late date under low grade conditions, and the late granites are mylonitised in the vicinity of thrusts.

2 North Greenland Fold Belt

This fold belt trends approximately E-W and is perpendicular to the East Greenland Caledonides. It consists of rocks ranging in age from Riphean to mid Silurian (Dawes 1971) and possibly up to the Devonian (Kerr 1967). The rocks are unconformably overlain by Carboniferous deposits which also form a cover both the East and the Northeast Greenland fold belts.

Peary Land shows the fullest sequence of Precambrian and Lower Palaeozoic rocks. An intensely folded Lower Palaeozoic turbidite sequence is seen in the north and east, whilst in central and south Peary Land unfolded shelf deposits are observed. Harland and Gayer (1972) state that between the two areas biohermal and biostromal carbonates are seen. In the southern part of the belt 1 km of arenaceous Riphean deposits are present and have been intruded by basic rocks prior to the deposition of a thin sequence (c. 110 m) of Vendian tillites and dolomites. Unconformably overlying these are a mixture of Cambrian to Silurian carbonates and clastics probably less than 3 km in thickness.

In Northwest Greenland a distinct sedimentary basin (Thule Basin) is observed and shows a different sequence of deposits to those in the main North Greenland fold belt. It consists of 1.3 km of Riphean rocks, the sequence being one of quartzites and sandstones overlain by shales (Dawes 1971). The shales are intruded by dolerites which are cautiously correlated with those which intrude the Upper Riphean rocks of southern Peary Land. Unconformably above the Riphean stratigraphy is 1 km of interbedded dolomite, sandstone and siltstone of Vendian age. It is noticeable that tillites are absent from this sequence; also noticeable is the lack of Palaeozoic rocks.

As noted above, the north part of the North Greenland Caledonides is strongly folded, and also metamorphosed. Higgins et al. (1981) suggest that the main deformation took place in Devonian to early Carboniferous times, appreciably later than events affecting East Greenland, and strictly not Caledonian. Trettin (1973) argues that it is an extension of the Franklinian geosyncline of Arctic Canada which shows spasmodic deformation from mid Ordovician times, but with a main event during the mid Devonian to Mississippian.

In North Greenland, Higgins et al. (1981) identify three co-axial fold phases trending E-W, with the intensity of tectonism and metamorphism greatest in the north. Major thrusting does not seem to be present. Unlike East Greenland, crystalline basement is not involved in the North Greenland fold belt, and granitic intrusions are lacking (Higgins et al. 1981).

The North Greenland fold belt is altogether very different to the East Greenland belts, and clearly does not show the same sedimentary or tectono-metamorphic history.

Spitsbergen

The geological evolution of different areas of Spitsbergen contrast with each other in many aspects of their pre-Carboniferous history. The adjoining nature of these areas seen today is therefore unlikely to have been in existence until the Carboniferous. The cause of their juxtaposition is considered by Harland (1978) to be due to late Devonian major sinistral strike slip movement. Harland (1978) discusses the evolution of ten areas of Svalbard, but for simplicity the region can be divided into four main terrains.

1 East Spitsbergen (areas east of the Billefjorden fault zone)

In East Nordaustlandet, high grade metamorphic rocks, including feldspathite and amphibolite are present. These have been regarded either as Archaean basement or equivalent to the lower Hecla Hoek sequence, found further to the west. Radiometric ages obtained are in the range 393–430 Ma due to Caledonian recrystallisation, and rather enigmatic ages of 581–618 Ma presumably represent partially reset isotopic systems. In west Nordaustlandet a 11–12 km sequence of argillaceous-arenaceous sediments (Harland 1978) passes up into 600 m of Vendian shales and tillites, followed by a Cambrian to ? Ordovician carbonate sequence 1 km thick.

In the westernmost part of East Spitsbergen (Ny Friesland) a sequence of Riphean to mid Ordovician rocks approximately 20 km thick is found. This is the type area of the Hecla Hoek sequence. The lower and middle Hecla Hoek are Riphean to basal Vendian in age, and make up 17–18 km of the sequence. The rocks of this sequence are largely quartzites, greywackes and shales, with volcanic rocks in the lower levels. The upper Hecla Hoek ranges in age from Vendian to middle Ordovician, and consists of 800 m of Vendian shales and tillite, followed by 1–2 km of lower Cambrian to mid-Ordovician dolomites and limestones which contain a rich Ordovician fauna. This Cambro-Ordovician sequence of shelf carbonates seems identical to the coeval strata of comparable age in central East Greenland, northwest Scotland and western Newfoundland (Swett 1981). It seems inescapable that the Cambro-Ordovician successions of these areas were developed on a once contiguous shelf on the northwest side of the Iapetus Ocean. Indeed the Riphean to Vendian histories of central East Greenland and East Spitsbergen show marked similarities.

2 Northwest Central, and Northwest Spitsbergen

At the base of the sequence in the northwest central area are highly tectonised high grade metamorphic rocks including eclogites and migmatites. Harland (1978) tentatively suggested them to be "Lower Hecla Hoek" equivalents, although radiometric dates give some enigmatic figures in the 1430–1541 Ma range as well as some Caledonian ages of 365–431 Ma. The remainder of the sequence is very different to that of East Spitsbergen. There is no thick Riphean to Ordovician sequence like the middle and upper Hecla Hoek, and following a period of orogeny probably at the end Ordovician/early Silurian, there was deposition of 1.5 km of late Silurian sandstones and conglomerates. Slight deformation of these rocks occurred (Haakonian event of Gee 1972) prior to the deposition of 7 km of continental sandstones and conglomerates of early and mid Devonian age. These deposits have experienced late Devonian (Svalbardian) deformation especially in the vicinity of the Billefjorden Fault Zone, along which there was major sinistral strike slip movement, the exact magnitude of which is unknown.
Apart from the Horneman granite (318–340 Ma), all rocks of the northwest corner of Spitsbergen are metamorphosed. They give ages of 375–450 (late Ordovician to mid Devonian), suggesting long or episodic metamorphism during the mid Palaeozoic, or possibly a prolonged cooling history.

3 West Spitsbergen

This area encompasses the ground from Kongsfjorden southward to the Bellsund area, and includes the island of Prins Karls Forland. The mainland area has an interrupted sequence of rocks of Riphean to Silurian age. The most intense deformation in the mid Palaeozoic gave rise to westward overthrusting with some accompanying metamorphism. However, a Tertiary orogenic episode with eastward overthrusting has produced a complex structural pattern such that Harland (1978) recognised five main tectonically separated sequences in the mainland area. The Vestgotabreen Formation contains blueschist metavolcanic rocks and yields a Caledonian date. Other units contain a mixture of clastic and carbonate deposits, and in the Comfortlessbreen Group, tillites are observed. Two of the other tectonically separate units contain Silurian fossils, and thus appear to have been deposited between the end Ordovician to Devonian phases registered in Northwest Spitsbergen.

Prins Karls Forland has an essentially similar sequence and history to the mainland of West Spitsbergen, although a less disrupted sequence upwards from the Vendian tillite is developed. Whilst west Spitsbergen preserves a Cambro-Ordovician sequence, this is not the thick carbonate shelf sequence that occurs in East Spitsbergen, but rather a mixture of interbedded sandstones, shales, conglomerates and carbonates. The West Spitsbergen sequence shows greater similarities with the North Greenland fold belt than with other Caledonian areas and may have originated to the north of Greenland. Late Devonian sinistral transcurrence may have juxtaposed East and West Spitsbergen (Harland and Gayer 1972).

4 Southernmost Spitsbergen

As with West Spitsbergen, this area was also affected by the Tertiary, West Spitsbergen Orogeny and therefore has an extra tectonic complication. Essentially the area consists of 1 km of Old Red Sandstone facies rocks lying unconformably above a very disrupted Riphean to Ordovician sequence. The breaks in this sequence are the consequence of spasmodic minor tectonic events. The main Riphean sequence consists of 3 km of amphibolites, feldspathites and various sediments including a tillite. It is overlain by 8 km of late Riphean and Vendian sediments including tillites and a greater quantity of dolomite and limestone. This sequence passes up into 800 m of lower Cambrian dolomite, limestone and shale, then 1.4 km of lower Ordovician limestone and dolomite, and clearly shows most similarity to the sequence of East Spitsbergen and East Greenland.

Tectonic Models

Early models relating the development of the Caledonian-Appalachian orogen to the evolution of an early Palaeozoic ocean (Wilson 1966, Dewey 1969) coincided with the formulation of plate tectonic theory (McKenzie and Parker 1967, Morgan 1968). These early models were simplistic and failed to account for many of the observed field relations. Not surprisingly there have been numerous plate models developed subsequently, each stressing the significance of a new piece of local data or drawing broad analogies with modern plate boundaries (eg. Phillips et al. 1976). In general, with the addition of more information, successive models have become more complex. This trend has led Dewey (1982) to speculate that the problems of producing an all embracing plate model for the orogen are almost insurmountable; particularly when one considers that the Caledonian-Appalachian orogen must have passed through stages comparable in complexity to modern plate boundaries, such as the western and eastern Pacific and the Mediterranean region. The probable 2000 km of late Palaeozoic sinistral transcurrence seemingly demanded by the palaeomagnetic data (Van der Voo et al. 1979) merely adds to the problem.

Nevertheless, it is clear from the scale factor (Dewey 1982) that any attempt at a realistic plate model must embrace a larger sector of the belt than any of the constituent portions into which it has been fragmented by Palaeozoic tectonics. Thus in this analysis we draw on as wide a spectrum of geological data as possible from the whole belt, and attempt to re-construct palaeotectonic maps at eight specific stages in the history of its evolution. It is clear that the result is unlikely to be more than an impression of the reality, for at each stage often even the relative positions of neighbouring terrains is unknown. Particularly during the final continent-continent collision, the effects of strike-slip transpressive movements will have shuffled smaller and even larger terrains in a way that will be hard to recognise from the preserved record. Bearing these reservations in mind, the eight maps presented are accompanied by a brief commentary indicating the significance of individual relationships and outlining some of the more important pieces of evidence used to infer particular plate configurations.

There are several major recurring considerations that have been used in constructing the models. Firstly, the presence in different terranes of benthonic faunas related to a common faunal province is taken to imply derivation from the same continental margin. Conversely, the presence of disparate faunas within neighbouring terranes suggests oceanic separation. Secondly, the presence of obducted ophiolitic sequences and island arc related volcanicity suggests the former presence of an ocean and thus the possibility of major lithospheric shortening by subduction of the intervening ocean. The polarity of subduction is controlled by directional chemical zoning of the arc volcanism and by the sense of obduction which is assumed to result from arc-continent or arc-arc collision, with obduction onto the descending plate (eg. Dewey and Bird 1970). Thirdly, the occurrence of an orogenic episode with associated plutonism, regional metamorphism and large-scale thrusting is taken to be the result of collisional tectonics involving two low density (buoyant) lithospheric plates. This may imply major continent-continent collision or the collision between a continent and a microcontinent-island arc complex, as in the case of the North American western cordillera (eg. Mattauer et al. 1983). In either case plutonism will be largely restricted to the overriding continental plate whilst thrusting will propagate away from the suture zone onto the underlying plate, as in the Himalayas (eg. Windley 1983).

Late Precambrian-Late Cambrian

Iapetus Rifting

Deposition in early Caledonian Precambrian sedimentary basins occurred in all the terrains that are related to the Caledonian continental margins. The oldest of these was developed on the northwest margin in the Arctic where deposition of the Eleonore Bay Supergroup of central East Greenland and the correlated Hecla Hoek Supergroup of East Svalbard may have commenced as early as 950 Ma (Harland and Gayer 1972). These sequences, with thick volcanic units in their lower parts, appear to have developed in an extensional basin upon thinned continental crust. On the southeast margin, the late Riphean sequences of the Baltic Cover thrust sheets in Scandinavia were developed in ensialic basins. Widespread tholeiitic dyke swarms with transitional MORB/within plate chemistry intrude into these basins in central and north Scandinavia and are thought to represent initial Iapetus rifting (Solyom et al. 1979, Gayer et al. in press) at ca. 640–700 Ma (Claesson and Roddick 1981). The approximately coeval alkali basaltic lavas of the Hedmark Group in southern Scandinavia may represent 'off axis' volcanicity associated with this rift event.

Further south there is more definite evidence for a late Precambrian ocean development although the timing of the initial rifting is uncertain, being possibly as old as 800 Ma (Strong 1974). Along the northwest margin in Scotland and the Appalachians a thick late Precambrian sequence developed at a passive continental margin on the eroded Proterozoic Grenvillian rocks (Hatcher this volume). On the southeast margin the late Precambrian clastic sediments of Southern Britain were deformed in the subduction related Mona Orogeny at ca. 600 Ma (Barber 1971, Thorpe 1979 and Gibbons in press), clearly indicating the existence of a mature ocean at this time.

Tectonic Analysis and Evolution of Related Oceans   149

It is tentatively suggested that Iapetus was initiated earlier in the south than the north, with rifting between the African and Laurentian cratons in the Appalachians occurring at ca. 800 Ma but not extending northwards between the Greenland and Baltic cratons until 700–640 Ma. Extensional basins in the north, however, developed over thinned continental crust prior to rifting.

Iapetus Spreading

Throughout latest Precambrian and much of Cambrian time, with the exception of the Monian orogeny in Anglesey, Iapetus appears to have had passive continental margins. The earliest development of destructive margins was in the mid Cambrian of northern-most Norway with the Finnmarkian Orogeny at 540 Ma (Sturt et al. 1978) (Fig. 5) and

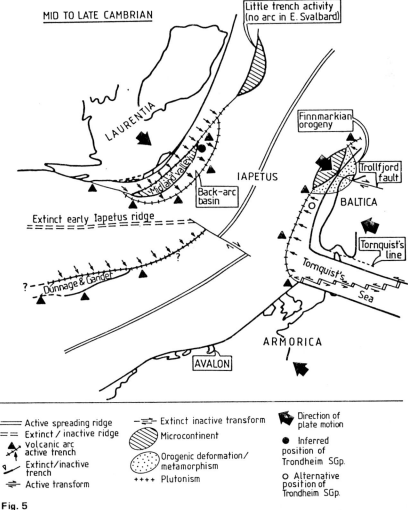

Fig. 5

Mid to Late Cambrian palaeotectonic map of the Caledonide-Appalachian orogen. See text for explanation.

in the latest Cambrian of Scotland and Northern Ireland with the Grampian orogeny at ca. 500 Ma (Lambert and McKerrow 1976). If we assume a modest single flank spreading rate of 2 cm yr^{-1}, Iapetus will have reached a width of ca. 10.000 km in the south, narrowing to ca. 6000 km in the north. This was sufficient to have produced very distinctive faunal provinces on its margins. It is also possible that the Cambro-Ordovician carbonate shelf of the northwestern Appalachians, Scotland, East Greenland and East Svalbard was developed in tropical latitudes, whereas the clastic shelf of the southeast margin may have lain significantly farther south, possibly in temperate latitudes (Spjeldnaes 1961, Cocks and Fortey 1982).

Marginal Island Arcs and Microcontinental Fragments

During late Cambrian time, oceanward subduction is indicated along the northwest margin of Iapetus (Fig. 5). The evidence for this comes partly from basic volcanic rocks and mélanges in the Dunnage terrane developed above ophiolite complexes, and indicating ensimatic island arcs. The island arc activity lasted from the late Cambrian to the early Ordovician. Immediately to the east of the Dunnage terrane lies the Gander terrane with included magmatic activity suggesting southeastward dipping subduction beneath a continental margin.

The precise geographic locations of these fragments of oceanic and continental crust during the Late Cambrian is uncertain. Mid Cambrian to early Ordovician faunas in the Dunnage terrane suggest 'Gondwana' affinities but endemic faunal elements are also present suggesting island communities (Dean, this volume). It would thus seem possible that the Gander terrane represents a microcontinental fragment separated from Gondwanaland and drifted across Iapetus during late Cambrian and early Ordovician ocean spreading. This would imply a second generation ocean spreading and the development of a ridge-transform-trench system within the Iapetus ocean (Fig. 5).

Along the southeast margin of Iapetus the main evidence for oceanward subduction comes from northernmost Norway where the Finnmarkian orogeny is thought to have resulted from the collision of the Baltic craton with a postulated microcontinental fragment possibly rifted from the Baltic margin early in the development of the Iapetus Ocean. Thrusting in the Finnmarkian orogeny was directed towards the southeast, across the Baltic foreland, implying that the Baltic craton lay on the underriding plate. The lack of subduction related magmatism within the Finnmarkian deformed zone, apart from the Seiland igneous province which may have originated from a deep mantle plume (Sturt in Gayer et al. in press), also suggests northwesterly directed subduction.

There is no direct evidence for the volcanic arc west of the main Baltic passive margin which is required by the early Ordovician phase of ophiolite obduction. Similarly the double zone of destructive margins southeast of Scotland is required to explain the early Ordovician collisional Grampian orogeny and Ballantrae ophiolite obduction.

Early Ordovician

Microcontinent-Continent Collision and Ophiolite Obduction

The Finnmarkian orogeny continued into earliest Ordovician time with the propagation of the foreland thrust sheets across the Baltic margin. At the same time the marginal ocean basins farther south, to the west of the Baltic craton, were closed by continued northwestward subduction resulting in arc-continent collision and obduction of ophiolites southeastwards across the Baltic margin (Fig. 6). There is some evidence from isotopic

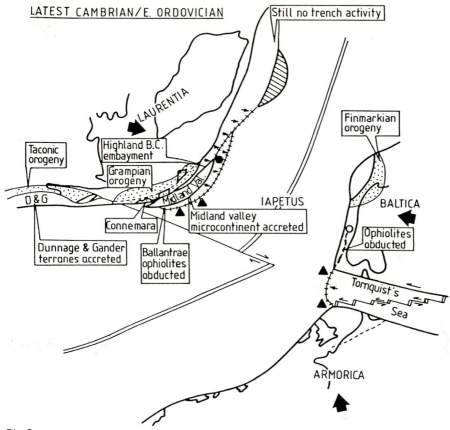

Fig. 6
Latest Cambrian to Early Ordovician palaeotectonic map of the Caledonide-Appalachian orogen. Symbols as in Figure 5.
Highland B.C. = Highland Boundary Complex. See text for explanation.

data that the Precambrian basement was partially reworked during this deformational event. This collision and obduction was not developed further south in southeastern Britain and sinistral transcurrence along the Tornquist Line is implied.

The main early Ordovician activity occurred along the northwest margin of Iapetus. In the western Appalachians, closure of the ocean west of the Dunnage and Gander terranes caused a collision between the Laurentian craton and the combined microcontinent and arc complex (Fig. 6). The resulting Taconic orogeny was produced along the lenght of the Appalachians with metamorphism in the Piedmont terrane and westward propagating thrusts across the carbonate platform. In the northern Appalachians, ophiolite obduction northwestwards across the miogeocline produced the Taconic allochthons. The asymmetry of the Taconic orogeny clearly indicates southeasterly directed subduction prior to and during collision.

To the northeast, the Grampian orogeny was also produced at this time by collision between the Laurentian craton and a microcontinent — the Midland Valley terrane (Fig. 6). The Highland Boundary Complex represents a fragment of oceanic lithosphere caught in the suture zone. The discrepancy in age between the Grampian orogeny and the younger Highland Boundary Complex could have resulted from late strike-slip adjustments associated with embayments and promontories along the microcontinental margin (Fig. 6). High grade regional metamorphism occurred in the Grampian Highlands with a small quantity of crustal anatexis. The northwestward directed thrusting in the Dalradian and Moine Supergroups suggests that, as in the Appalachians, the Laurentian craton lay on the underriding plate. The principal difference between the Grampian and Taconic orogenies lies in the derivation of the colliding microcontinents. The Midland Valley terrane contains early Ordovician faunas of Laurentian affinity and was separated from the Laurentian craton by marginal ocean spreading during early Iapetus rifting, whereas the Gander/Dunnage terranes existed as a mid-ocean microcontinent, or were derived from the southeast margin of Iapetus.

In western Ireland, the Connemara Dalradian lies south of the westward continuation of the Highland Boundary Fault and is separated from the main outcrop of Dalradian rocks by the early to middle Ordovician succession of the South Mayo trough (Dewey 1963). The infill of the trough unconformably overlies and obscures the contact between the two areas of Dalradian. Similarly the contact between the South Mayo trough and the Connemara Dalradian is obscured by unconformably overlying Upper Llandovery rocks. Various explanations have been suggested for the presence of the Connemara Dalradian south of the Highland Boundary Fault (eg. Leake et al. 1983) but none is entirely satisfactory. We suggest that during the final stages of collisional orogeny between Laurentia and the Midland Valley microcontinent, dextral strike-slip displacement shifted the Connemara block westwards from a position north of the suture. A bend in this transcurrent fault could have resulted in a pull-apart basin within which was deposited the post-Dalradian sequence of the South Mayo trough. Closure of the pull-apart basin would have resulted in both the deformation of the infill of the trough and the southward directed thrusting of the Connemara Dalradian observed by Leake et al. (1983).

At about the same time in the late early Ordovician, the Ballantrae ophiolite was obducted northwestwards across the southern margin of the Midland Valley terrane as a result of collision with an arc that lay to the southeast.

The original location of the Trondheim Supergroup of central Norway is controversial. The faunas indicate a Laurentian affinity (eg. Bruton and Bockelie 1980), but sedimentological evidence suggests development on the western edge of a continental platform which Ryan et al. (1980) argued represented the Baltic Craton. Roberts (1981), supporting Ryan et al. (1980), showed that the included Arenigian/Llanvirnian volcanic rocks were developed in an arc environment on transitional oceanic/continental crust above a southeasterly dipping subduction zone. An alternative interpretation is that the Trondheim Supergroup may have been developed on the western margin of a northward continuation of the Midland Valley microcontinent. This would satisfy both faunal and sedimentological requirements whilst southeasterly directed subduction would be consistent with that postulated for the Scottish area. However, in this region, instead of continental collision resulting in the equivalent of the Grampian orogeny, the Trondheim Supergroup suffered only mild orogenic deformation (Ryan et al. 1980) perhaps as a result of lying in a major embayment on the edge of the Midland Valley microcontinent. Figs. 5–9 show the two alternative Locations for the Trondheim Supergroup.

Mid-Late Ordovician

Fringing Island Arcs

At this time a major change took place with the gradual closure of Iapetus. This is indicated by the loss of provinciality in the mid and late Ordovician faunas to either side of the belt and also by the development of marginally directed subduction zones to either side of the ocean (Fig. 7). The change may reflect a combination of slowing down of Iapetus spreading and converging motions of the bordering continental plates. Subduction is most clearly developed along the southeast margin, where ensimatic island arcs were developed in Llanvirn times along the entire Baltic margin from Troms southwards. These are represented in the volcanic sequences of the Balsfjord Group (Gayer et al. 1984) and the Melkefjell amphibolite formation (Barker 1984) of Troms and the higher tectonic units of the Köli Supergroup (Stephens 1980) in Central Scandinavia. Associated with the trench/arc activity back arc extension and arc splitting occurred with the development of the Skibotn (Gayer et al. 1984) and Melkefjell (Barker 1984) amphibolites in Troms and the Sulitjelma amphibolite (Boyle et al. 1979) of the lower tectonic units of the Köli Supergroup.

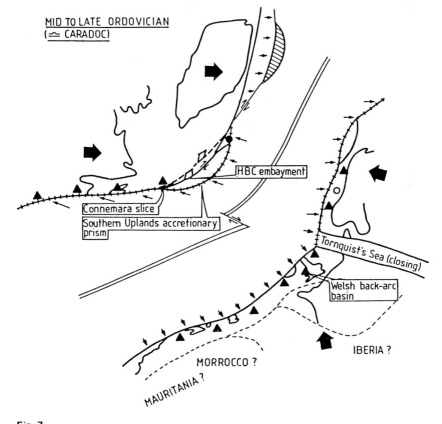

Fig. 7
Mid to Late Ordovician (Caradoc) palaeotectonic map of the Caledonide-Appalachian orogen. Symbols as in Figure 5.
H.B.C. = Highland Boundary Complex. See text for explanation.

In the paratectonic Caledonides of southeastern Britain a similarly southeasterly directed subduction occurred in Llandeilo times with the development of the Lake District volcanic arc (Fitton and Hughes 1970). The Welsh basin appears to have been in part a back arc extensional basin (Kokelaar et al. in press) but developed entirely in an ensialic environment. A possible element of transtensional tectonics may also have been involved with dextral strike-slip movements on many of the controlling faults (Woodcock in press). Much of the Ordovician volcanicity within the Welsh basin appears to have been associated with these faults.

The analogous northwesterly directed subduction along the northern margins of Iapetus is not so clearly documented. The most obvious evidence comes from the Southern Uplands of Scotland, along the southeast margin of the Midland Valley terrain, where a major accretionary prism developed from Llandeilo through Llandovery times. The associated volcanic arc is not well developed and may have been tectonically concealed when the accretionary prism was thrust northwestwards during final Iapetus closure, or may have become separated from the accretionary prism during strike-slip movement along the S.U.F. (see Gibbons and Gayer, this volume).

Early Silurian

The main activity was associated with subduction of Iapetus along the northwest margin of the ocean (Fig. 8). The Southern Uplands accretionary prism had developed to form an outer arc (Cockburnland) which supplied sediment northwards into the Midland Valley basin. Along the southeast margin of the ocean, to the north of the Tornquist

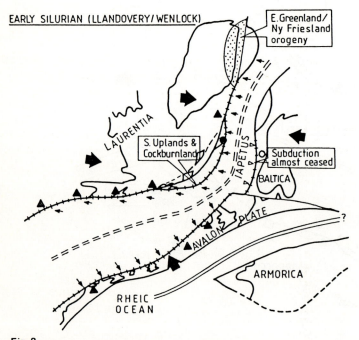

Fig. 8
Early Silurian (Llandovery-Wenlock) palaeotectonic map of the Caledonide-Appalachian orogen. Symbols as in Figure 5. See text for explanation.

Line, subduction had virtually ceased with the exception of arc volcanics in the Köli Supergroup, and it may be that ridge activity had also ceased. To the south, the Rheic ocean started to open, separating the Avalon Plate (Piqué 1981) from Gondwanaland. There is a possibility that the Baltic Plate moved northwest more rapidly resulting in continued sinistral slip along the Tornquist line and leading to closure of the northern sector of Iapetus.

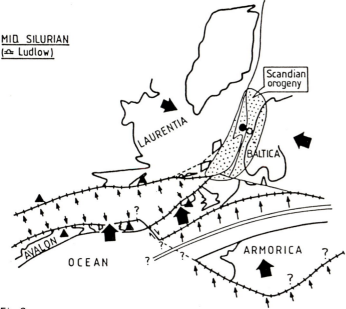

Fig. 9

Mid Silurian (Ludlow) palaeotectonic map of the Caledonide-Appalachian orogen. Symbols as in Figure 5. See text for explanation.

Mid Silurian

Baltica collided with Laurentia giving rise to the Scandian orogeny (Fig. 9). The suture between the two continental plates is concealed by overthrusting of the Trondheim nappe complex from northwest of the suture onto the Baltic plate. The Trondheim nappe complex forms part of the upper allochthon in a sequence of southeastward propagating thrust sheets, suggesting that Baltica was overridden by Laurentia. The higher tectonic units have undergone high grade reginal metamorphism but only minimal magmatic activity, supporting the northwesterly polarity of subduction prior to and during collision. The British area and the Appalachians did not experience major deformation at this time since the Avalon plate had not yet closed with Laurentia (Fig. 9).

Late Silurian — Earliest Devonian

The final stages of the Scandian orogeny were produced by continued subduction of the Baltic plate beneath Laurentia, giving rise to eclogite facies metamorphism in western Baltica with higher pressure and temperature conditions westwards. This event probably

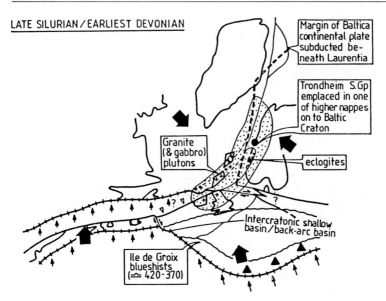

Fig. 10
Late Silurian to Earliest Devonian palaeotectonic map of the Caledonide-Appalachian orogen. Symbols as in Figure 5. See text for explanation.

coincides with the final emplacement of the Trondheim nappe complex over the Baltic margin (Fig. 10).

The principal development at this time was the collision between the British margin of the Avalon plate with Laurentia, producing low grade regional metamorphism and upright folding (Fig. 10). The absence of intense deformation and metamorphism indicates that Britain lay in an embayment along the colliding continental margins. The Newfoundland/ Appalachian sector of the Avalon plate was separated from Laurentia by the remnant of Iapetus. Faunal provinciality indicates that the Rheic ocean had probably reached its maximum width (Cocks and Fortey 1982).

Mid Devonian

The main part of the Avalon plate was accreted to the Laurentian plate to produce the collisional Acadian Orogeny from Newfoundland southwards through the Appalachians (Fig. 11). High temperature and relatively high pressure metamorphism and plutonic magmatism were largely restricted to central and southern New England which may have been the site of a promontary during collision. The effects of the Acadian deformation were greatest in the central and eastern Appalachians and generally diminished westwards in the Humber-Zone. The result of this collision was to produce a major continental mass — the Old Red Continent or Laurasia. The Rheic ocean to the south, separating this continent from Gondwanaland, was already experiencing subduction along its northern margin with the development of the Ligerian Arc and the Rheno-Hercynian back arc basin (Leeder 1976). Following collision the plate motion changed to one of sinistral transcurrence and major strike-slip faults, such as the Billefjorden Fault Zone in Svalbard (Harland 1979), the Great Glen Fault in Scotland (Kennedy 1946) and the Dover Fault in Newfoundland, were initiated. Associated with this transcurrent regime was the devel-

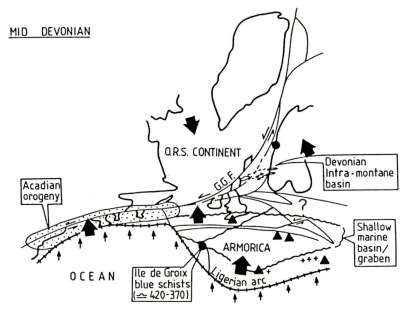

Fig. 11
Mid Denovian palaeotectonic map of the Caledonide-Appalachian orogen. Symbols as in Figure 5.
G.G.F. = Great Glen Fault; O.R.S. = Old Red Sandstone. See text for explanation.

opment of a number of molassic basins infilled with Old Red Sandstone facies sediments. These are well developed in Central Svalbard, along the west coast of Norway (Nilsen 1968) and Northern Scotland (Fig. 11). The basins are generally fault controlled and they are thought to have resulted from transtensional tectonics related to the principal sinistral faults (Roberts 1983).

End Devonian

The sinistral strike-slip movements, initiated in early to mid Silurian time, continued but the overall regime altered to one of transpression with the development of folding and thrusting within the earlier pull-apart basins. This phase of deformation — the Svalbardian — represents a change in plate motions probably resulting from closure of the back arc basin along the northern margin of the Rheic ocean and the accretion of the Meguma terrain onto the Laurasian margin.
The 1 500—2 500 km of sinistral strike-slip movement indicated by palaeomagnetic data (Storevedt 1973, Morris 1976, Kent and Opdyke 1979, Vander Voo et al. 1979) was accomplished over the period from early Devonian through early Carboniferous by smaller movements along a number of broadly parallel faults. This resulted in differential displacements of individual terranes along the mega-shear. Thus terranes such as Meguma, Avalon and NW Scotland were transported to their present juxta-positions and Greenland, Scandinavia and Svalbard took up their relative positions pre-North Atlantic spreading (Fig. 12).

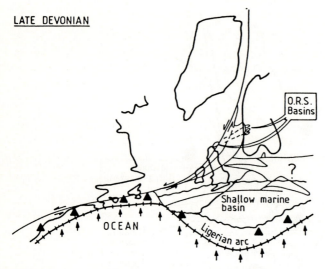

Fig. 12
Late Devonian palaeotectonic map of the Caledonide-Appalachian orogen. Symbols as in Figure 5.
O.R.S. = Old Red Sandstone. See text for explanation.

Conclusions

The Caledonide-Appalachian orogen is constructed from a large number of terranes, each possessing distinctive depositional and/or tectono-metamorphic histories and whose present geometry and juxtaposition has resulted from a combination of compressional and transcurrent movements during the closure of the Iapetus and Rheic Oceans.
The following terranes have been identified and described:

Southeast Iapetus Margin

Baltic Terrane — Caledonian deposition occurred from late Riphean through Devonian on the northwest edge of the Balto-Scandian Precambrian Shield. Deformation was in the Finnmarkian event (Mid Cambrian-Early Ordovician) and in the Scandian event (Mid Silurian).
Two subterrains have been recognised, separated by thrust boundaries:
a) Baltic Cover Thrust Sheets
b) Crystalline Basement Thrust Sheets
Avalon Terrane — Late Precambrian sediments, volcanic rocks and plutonic intrusions overlain unconformably by Cambrian and younger marine sediments containing 'Gondwanaland' province faunas characterise this terrane. Deformation occurred in the Mid Devonian, Acadian orogeny. Geographically separated parts of the terrane are found in southern Britain, southeast Newfoundland, Nova Scotia and the eastern S. Appalachians.
The *Monian Terrane* of northwest Wales is separated by major vertical shear zones from the main *southern Britain Terrane* to the southeast and represents a late Precambrian oceanic development (Gibbons and Gayer, this volume).

Intra-Iapetus

Scandian Oceanic Terrane — Characterised by ophiolitic slices obducted onto the Baltic margin during the early Ordovician. The overlying calc-alkaline volcanic rocks and marine sediments were developed from early Ordovician through early Silurian. The lower thrust sheets contain Baltic province faunas, whereas those of the upper Trondheim nappe complex are of Laurentian province affinity. The main orogenic deformation occurred during the mid Silurian.

Southern Scotland Terrane — Consistens of Ordovician through Silurian marine sediments, containing Laurentian province faunas, deposited on Precambrian crystalline basement. Obducted ophiolites are present in subterranes. Deformation occurred locally during the mid Arenig associated with ophiolite emplacement, but generally in Late Silurian/early Devonian.

Four subterranes can be distinguished:

a) *Southern Uplands terrane:* Consists of a mid Ordovician through mid Silurian accretionary prism, thrust northwards over crystalline basement. Deformation occurred in late Silurian/early Devonian, followed by post-tectonic granitic intrusions.

b) *Midland Valley terrane:* Crystalline basement occurs beneath a thick late Palaeozoic cover. Oldest exposed sediments are early Silurian with subduction generated late Silurian/early Devonian volcanic rocks. Deformation was developed in mid Devonian. The obducted Ballantrae ophiolite may represent a separate subterrane.

c) *Highland Boundary terrane:* Ophiolite and associated early Ordovician volcanic rocks and sediments were deformed during the mid Arenig. Separated from the neighbouring terranes by the Highland Boundary fault zone.

d) *Connemara terrane:* A microterrane involving the Connemara Dalradian and the south Mayo trough associated with movements along the Highland Boundary fault zone.

Central Appalachian Terrane — Characterised by early Palaeozoic ophiolites, mafic volcanic rocks, mélanges and marine sediments, containing endemic island faunas but related to the 'Gondwanaland' province. Precambrian continental basement gneisses occur in the southeast, and deformation was in both the Taconic (early mid Ordovician) and the Acadian (mid Devonian) orogenies.

Two principal subterranes are recognised in the NE Appalachians:

a) *Gander terrane:* continental basement with high grade regional metamorphism.

b) *Dunnage terrane:* ophiolitic basement representing Iapetus ocean crust and mantle, with low grade metamorphism.

Northwest Iapetus Margin

The province is characterised by a late Riphean through early Palaeozoic miogeoclinal sequence associated with the passive continental margin of Laurentia. A particularly distinctive Cambro-Ordovician carbonate platform sequence is developed containing typical Laurentian province faunas.

East Greenland and East Svalbard Terranes — These are two related terranes separated by mid Devonian sinistral transcurrence along the extension of the Billefjord fault zone. The terranes are characterised by a very thick (19 km) late Riphean through mid Ordovician sequence, deformed during ? early Silurian.

Trondheim Terrane — This complex terrane consists of the highest nappes of the Scandian allochthon. Proterozoic basement gneisses, re-metamosphosed and thrust southeastwards during the Scandian orogeny (mid Silurian), have been derived from west of the *Scandian Oceanic Terrane*. Early-mid Ordovician faunas associated with calc-alkaline volcanic rocks and ophiolites in the Trondheim nappe indicate Laurentian province affinities but oceanward of the Laurentian margin. Thus this terrane originated from the northwest Iapetus margin, south of the East Greenland terrane but north of the Northern Scotland terrane.

Northern Scotland Terrane — This terrane comprises the orthotectonic British Caledonides in which the Moine and Dalradian Supergroups were deformed during the Grampian orogeny (latest Cambrian-early Ordovician). Propagation of thrusts northwestwards continued into the mid/late Silurian. Mid Devonian transcurrent movement along the Great Glen fault has juxtaposed two subterranes:

a) *Grampian Highland terrane:* Characterised by high grade regional metamorphism of the Dalradian Supergroup deposited on Grenvillian? basement.

b) *Northern Highlands terrane:* Consisting of the late Proterozoic (Grenvillian) Moine Supergroup reactivated and thrust northwestwards across the carbonate platform Northwest Foreland Terrane during mid Ordovician — late Silurian.

c) *Northwestern Foreland terrain* consisting of Archaean (Lewisian) basement, unconformably overlain by Proterozoic (Torridonian) clastics. These are in turn overlain by a Cambro-Ordovician shelf sequence, partly imbricated by early Ordovician (Grampian) thrusting.

Humber Terrane — A miogeoclinal late Riphean through early Palaeozoic sequence deposited on Grenvillian crystalline basement. Deformation occurred in the Taconic orogeny (early-mid Ordovician) and in the Acadian orogeny (Mid Devonian). Three subterranes are recognised representing thrust telescoping of the original northwest Iapetus continental shelf, slope and rise environments.

a) *Appalachian Miogeocline:* Low grade metamorphic continental shelf and slope sequence.

b) *Taconic allochthons:* This composite terrane is mainly developed in the northern and central Appalachians, with the lowest thrust sheets preserving the continental slope and rise, and the highest thrust sheets the oceanic crust and mantle.

c) *Piedmont terrane:* Has its largest development in the southern Appalachians and consists of high grade metamorphosed clastic rocks representing original continental slope and rise.

The Palaeotectonic maps (Figs. 5—12) represent an attempt at repositioning these terranes at eight consecutive stages in the evolution of the belt. In general the closer a terrane to the Iapetus suture the greater the uncertainty of its original location. It is hoped that future research will serve to reduce this uncertainty so as to define more precisely the accretionary history of the orogen.

Acknowledgements

We are grateful for valuable discussions with many colleagues, in particular Professor Bill Dean and Dr Wes Gibbons and we are grateful to Mrs Anthea Dunkley for draughting the plate tectonic maps.

References

Anderton, R. (1982): Dalradian deposition and the late Precambrian-Cambrian history of the N Atlantic region: a review of the early evolution of the Iapetus Ocean. J. geol. Soc. London 139, 421—31.

Baker, J.W. (1971): The Proterozoic history of southern Britain. Proc. Geol. Ass. 82, 249—66.

Banks, N. L., Hobday, D. K., Reading, H. G. and Taylor, P. N. (1974): Stratigraphy of the Late Precambrian 'Older Sandstone Series' of the Varangerfjord area, Finnmark. Norges geol. Unders. 303, 1–16.
Barker, A. J. (1984): The geology between Gratangenfjord and Salangsdalen, S. Troms, Norway and its regional significance. Unpublished Ph. D. Thesis, University of Wales 419pp.
Bassett, M. G., Cherns, L. and Karis, L. (1982): The Röde Formation, Early Old Red Sandstone facies in the Silurian of Jämtland, Sweden. Sveriges geol. Unders. C793, 24 pp.
Beckinsale, R. D., Reading, H. G. and Rex, D. C. (1976): Potassium-argon ages for basic dykes from East Finnmark: stratigraphical and structural implications. Scott. J. Geol. 12, 51–65.
Bell, K., Blenkinsop, J. and Strong, D. F. (1977): The geochronology of some granitic bodies from eastern Newfoundland and its bearing on Appalachian evolution. Can. J. Earth Sci. 14, 456–76.
Blackwood, R. F. (1978): Northeastern Gander Zone. In: Gibbons, R. V. (edit.), Report of activities for 1977: Newfoundland Department of Mines and Energy, Mineral Development Division. Report 78-1, 72–9.
Boyle, A. P., Griffiths, A. J. and Mason, R. (1979): Stratigraphical inversion in the Sulitjelma Area, Central Scandinavian Caledonides. Geol. Mag. 116, 393–402.
Brewer, J. A., Brook, M. and Powell, D. (1979): Dating of the tectono-metamorphic history of the southwestern Moine, Scotland. In: Harris, A. L., Holland, C. H. and Leake, B. E. (eds.). The Caledonides of the British Isles – reviewed. Geol. Soc. London, Spec. Pub. 10, pp 129–137.
Brewer, J. A., Matthews, D. H., Warner, M. R., Hall, J., Smythe, D. K. and Whittington, R. J. (1983): BIRPS deep seismic reflection studies of the British Caledonides. Nature 305, 206–10.
Brook, M., Powell, D. and Brewer, M. S. (1976): Grenville age for rocks in the Moine of north-western Scotland. Nature, London 260, 515–17.
Brown, G. C. (1979): Geochemical and geophysical constraints on the origin and evolution of Caledonian granites. In: Harris, A. L., Holland, C. H. and Leake, B. E. (eds.). The Caledonides of the British Isles – reviewed. Geol. Soc. London. pp 645–651.
Bruton, D. L. and Bockelie, J. F. (1980): Geology and Paleontology of the Hølonda area, western Norway – a fragment of North America? Proc. IGCP Cal. Orogen Symp., Virginia Poly. Inst. Memoir 2, 41–47.
Bruton, D. L. and Lindström, M. (1981): The Ordovician of Scandinavia. Terra Cognita 1, 37.
Bryhni, I. and Andreasson, P. G. (1981): Metamorphism in the Scandinavian Caledonides. Terra Cognita 1, 37.
Bursnall, J. T. and De Wit, M. J. (1975): Timing and development of the orthotectonic zone in the Appalachian orogen of northwest Newfoundland. Can. J. Earth Sci. 12, 1712–22.
Church, W. E. (1969): Metamorphic rocks of the Burlington Peninsula and adjoining areal of Newfoundland, and their bearing on continental drift in North Atlantic. In: Kay, M. (ed.) North Atlantic – geology and continental drift – a symposium. Mem. Am. Ass. Petrol. Geol. 12, 212–35.
Cleasson, S. and Roddick, J. C. (1981): The age of the Ottfjället dolerites, Swedish Caledonides: New evidence from $^{40}Ar/^{39}Ar$ spectra. Terra Cognita 1, 40.
Cocks, L. R. M. and Fortey, R. A. (1982): Faunal evidence for oceanic separation in the Palaeozoic of Britain. J. geol. Soc. London 139, 465–478.
Coward, M. P. and Siddans, A. W. B. (1979): The tectonic evolution of the Welsh Caledonides. In: Harris, A. L., Holland, C. H. and Leake, B. E. (eds.). The Caledonides of the British Isles – reviewed. Geol. Soc. London Spec. Pub. 10, pp 187–198.
Curry, G. B., Ingham, J. K., Bluck, B. J. and Williams, A. (1982): The significance of a reliable Ordovician age for some Highland Border rocks in Central Scotland. J. geol. Soc. London 139, 451–54.
Dalmeyer, R. D. and Williams, H. (1975): $^{40}Ar/^{39}Ar$ ages from the Bay of Islands metamorphic aureole: their bearing on timing of the Ordovician ophiolite obduction. Can. J. Earth Sci. 12, 1685–90.
Dawes, P. R. (1971): The North Greenland fold belt and environs. Dansk. Geol. Foren. Medd. 20, 197–239.
Dewey, J. F. (1963): The Lower Palaeozoic stratigraphy of central Murrisk, County Mayo, Ireland, and the evolution of the South Mayo trough. Quart. J. geol. Soc. London 119, 313–44.
Dewey, J. F. (1969): The evolution of the Appalachian/Caledonian orogen. Nature, London 222, 124–9.
Dewey, J. F. (1982): Plate tectonics and the evolution of the British Isles. J. geol. Soc. London 139, 371–412.
Dewey, J. F. and Bird, J. M. (1970): Mountain belts and the new global tectonics. J. geophys. Res. 75, 2625–47.
Downie, C., Lister, T. R., Harris, A. L. and Fettes, D. J. (1971): A palynological investigation of the Dalradian rocks of Scotland. Rep. Inst. geol. Sci. London, 71/9.

Fitch, F. J., Miller, J. A., Evans, A. L., Grasty, R. L. and Meneisy, M. Y. (1969): Isotope age determinations on rocks from Wales and the Welsh Borders. In: Wood, A. (ed.). The Precambrian and Lower Palaeozoic rocks in Wales. University of Wales Press.
Fitton, D. J. and Hughes, D. J. (1970): Volcanism and plate tectonics in the British Ordovician. Earth Planet. Sci. Lett. 8, 223−8.
Føyn, S. (1967): Dividal-gruppen ("Hyolithus-sonen") i Finnmark og dens forhold til de eokambrisk-kambriske formasjoner. Norges geol. Unders. 249, 1−84.
Furnes, H., Roberts, D., Sturt, B. A., Thon, A. and Gale, G. H. (1980): Ophiolite fragments in the Scandinavian Caledonides. In: Ophiolites. Proceedings of International Ophiolite Symposium. Cyprus Geol. Surv. Dept. pp 582−600.
Gale, G. H. and Roberts, D. (1974): Trace element geochemistry of Norwegian Lower Palaeozoic basic volcanics and its tectonic implication. Earth Planet. Sci. Let. 22, 380−390.
Gaudette, H. E. (1981): Zircon isotopic age from the Union ultramafic complex, Maine. Can. J. Earth Sci. 18, 405−9.
Gayer, R. A., Humphreys, R. J., Chapman, T. J. and Binns, R. E. (in press): Tectonic modelling of the Finnmark and Troms Caledonides based on high-level igneous rock geochemistry. In: Gee, D. G. and Sturt, B. A. (eds.) The Caledonide orogen − Scandinavia and related areas. Wiley.
Gayer, R. A., Gee, D. G., Harland, W. B., Miller, J. A., Spall, H. R., Wallis, R. H. and Winsnes, T. S. (1966): Radiometric age determinations on rocks from Spitsbergen. Skr. Norsk Polarinst. 137, Oslo 39 pp.
Gee, D. G. (1972): Late Caledonian (Haakonian) movements in northern Spitsbergen. Norsk Polarinst. Arbok (1970), 50−54.
Gee, D. G. (1975): A tectonic model for the central part of the Scandinavian Caledonides. Amer. J. Sci. 275A, 468−515.
Gibbons, W. (in press): The Precambrian basement to England and Wales. Proc. Geol. Ass. London.
Hall, J., Powell, D. W., Warner, M. R., El-Isa, Z. H. M., Adesanya, O. and Bluck, B. J. (1983): Seismological evidence for shallow crystalline basement in the Southern Uplands of Scotland. Nature, London 305, 418−20.
Haller, J. (1970): The tectonic map of Greenland. Meddr. Grønland. 1971. The Geology of the East Greenland Caledonides. Wiley.
Halliday, A. N., Aftalion, M., van Breeman, O. and Jocelyn, J. (1979): Petrogenetic significance of Rb-Sr and U-Pb isotopic systems in the 400 Ma old British Isles granitoids and their hosts. In: Harris, A. L., Holland, C. H. and Leake, B. E. (eds.). The Caledonides of the British Isles − reviewed. Geol. Soc. London, Spec. Pub. 10, pp 653−61.
Hampton, C. M. and Taylor, P. N. (1983): The age and nature of the basement of southern Britain: evidence from Sr and Pb isotopes in granites. J. geol. Soc. London 140, 499−509.
Harland, W. B. (1978): The Caledonides of Svalbard. Geol. Surv. Canada Pap 78−13, 3−11.
Harland, W. B. (1979): A review of major faults in Svalbard. In: Proceedings of Conference VIII − Analysis of actual fault zones in bedrock, 1−5 April 1979, U.S. Department of the Interior, Geological Survey, Office of Earthquake Studies, Menlo Park, Ca. Openfile report 79−1239, pp 157−80.
Harland, W. B. and Gayer, R. A. (1972): The Arctic Caledonides and earlier oceans. Geol. Mag. 109, 289−314.
Harris, A. L. and Pitcher, W. S. (1975): The Dalradian Supergroup. In: Harris, A. L. et al. (eds.). A correlation of Precambrian rocks in the British Isles. Spec. Rep. geol. Soc. London 6, 52−75.
Hatcher, R. D.: The tectonic evolution of the Appalachian orogen. This volume.
Henriksen, N. (1978): East Greenland Caledonian Fold Belt. Geol. Surv. Canada Paper 78−13, 105−110.
Higgins, A. K. and Phillips, W. E. A. (1979): East Greenland Caledonides − an extension of the British Caledonides. In: Harris, A. L. et al. (eds.). The Caledonides of the British Isles − reviewed. Geol. Soc. London. Spec. Pub. 10, pp 19−32.
Higgins, A. K., Soper, N. J. and Friderichsen, J. D. (1981): The North Greenland fold belt − a review. Terra Cognita 1, 52.
Hossack, J. R. and Cooper, M. (in press). Collisional tectonics in the Scandinavian Caledonides. J. geol. Soc. London.
Johnson, H. D., Levell, B. K. and Siedlecki, S. (1978): Late Precambrian sedimentary rocks in East Finnmark, north Norway and their relationship to the Trollfjord-Komagelv fault. J. geol. Soc. London 135, 517−533.
Kennedy, M. J. (1976): Southeastern margin of the southeastern Appalachians: Late Precambrian orogeny on a continental margin. Bull. geol. Soc. Am. 87, 1317−25.

Kent, D. V. and Opdyke, N. D. (1979): The early Carboniferous palaeomagnetic field of North America and its bearing on tectonics of the Northern Appalachians. Earth Planet. Sci. Lett. 44, 365–72.
Kent, D. V. and Opdyke, N. D. (1980): Palaeomagnetism of Siluro-Devonian rocks from eastern Maine. Can J. Earth Sci. 17, 1653–65.
Kerr, J. W. (1967): Stratigraphy of central and eastern Ellesmere Island, Arctic Canada. Part I. Proterozoic and Cambrian. Pap. geol. Surv. Can. 67–27(1), 63pp.
Kokelaar, B. P., Howells, M. F., Bevins, R. E., Roach, R. A. and Dunkley, P. N. (in press). The Ordovician marginal basin of Wales. In: Kokelaar, B. P. and Howells, M. F. (eds.). Volcanic and associated sedimentary and tectonic processes in modern and ancient marginal basins. Spec. Pub. geol. Soc. London.
Lambert, R. St. J. and McKerrow, W. S. (1976): The Grampian Orogeny. Scott. J. Geol. 12, 271–92.
Lambert, R. St. J. and Rex, D. C. (1966): Isotopic ages of minerals from the Precambrian complex of the Malverns. Nature, London 209, 605–6.
Lamont, A. and Lindström, M. (1957): Arenigian and Llandeilian cherts identified in the Southern Uplands of Scotland by means of conodonts, etc. Trans. Edinburgh geol. Soc. 17, 60–70.
Larsson, K. (1973): The Lower Viruan in the autochthonous Ordovician sequence of Jämtland. Sveriges geol. Unders. C683, 82pp.
Leake, B. E. (1978): Geol. J. Spec. Iss. 10, 221–248.
Leake, B. E., Tanner, P. W. G., Singh, D. and Halliday, A. N. (1983): Major southward thrusting of the Dalradian rocks of Connemara, western Ireland. Nature, London 305, 210–213.
Leeder, M. R. (1976): Sedimentary facies and the origins of basin subsidence along the northern margin of the supposed Hercynian Ocean. Tectonophysics 36, 167–79.
McKenzie, D. P. and Parker, R. L. (1967): The North Pacific: an example of tectonics on a sphere. Nature, London 216, 1276–80.
McKerrow, W. S., Leggett, J. K. and Eales, M. H. (1977): Imbricate thrust model of the Southern Uplands of Scotland. Nature, London 267, 237–9.
Mattauer, M., Collot, B. and Vanden Driessche, J. (1983): Alpine model for the internal metamorphic zones of the North American Cordillera. Geology 11, 11–15.
Max, M. D. (1975): Precambrian rocks of south-east Ireland. In: Harris, A. L. et al. (eds.). A correlation of Precambrian rocks in the British Isles. Spec. Rep. geol. Soc. London 6, 96–101.
Meneisy, M. Y. and Miller, J. A. (1963): A geochronological study of the crystalline rocks of Charnwood Forest, England. Geol. Mag. 100, 507–23.
Morgan, W. J. (1968): Rises, trenches, great faults, and crustal blocks. J. Geophysical Res. 73, 1959–82.
Morris, W. A. (1976): Transcurrent motion determined palaeomagnetically in the North Appalachians and Caledonides and the Acadian orogeny. Can. J. Earth Sci. 13, 1236–43.
Neuman, R. B. (1972): Brachiopods of Early Ordovician volcanic islands. 24th Int. Geol. Congr. Montreal Sec. 7, 297–302.
Nilsen, T. H. (1968): The relationship of sedimentation to tectonics in the Solund Devonian district of southwestern Norway. Norges geol. Unders. 259, 10pp.
Nystuen, J. P. (1982): Late Proterozoic basin evolution on the Baltoscandian Craton: the Hedmark Group, Southern Norway. Norges geol. Unders. 375, 1–74.
Phillips, W.E.A., Stillman, C. J. and Murray, T. (1976): A Caledonian plate tectonic model. J. geol. Soc. London 132, 579–609.
Piasecki, M.A.J. and van Breeman, O. (1979): The 'Central Highland Granulites': cover-basement tectonics in the Moine. In: Harris, A. L. et al. (eds.). The Caledonides of the British Isles — reviewed. Geol. Soc. London Spec. Pub. 10, 139–44.
Pidgeon, R. T. (1969): Zircon U-Pb ages from the Galway granite and the Dalradian Connemara, Ireland. Scott. J. Geol. 5, 375–92.
Piqué, A. (1981): Northwestern Africa and the Avalonian plate: Relations during late Precambrian and Late Palaeozoic time. Geology 9, 319–22.
Powell, D. (1974): Stratigraphy and structure of the western Moine and the problem of Moine orogenesis. J. geol. Soc. London 130, 575–93.
Powell, D., Brook, M. and Baird, A. W. (1983): Structural dating of a Precambrian pegmatite in Moine rocks of northern Scotland and its bearing on the status of the 'Morarian Orogeny'. J. geol. Soc. London 140, 813–24.
Pringle, I. R. (1973): Rb-Sr age determinations of shales associated with the Varanger ice age. Geol. Mag. 109, 465–72.
Reading, H. G. (1965): Eocambrian and Lower Palaeozoic geology of the Digermul Peninsula, Tanafjord, Finnmark. Norges geol. Unders. 234, 167–191.

Richardson, S.W. and Powell, R. (1976): Thermal causes of the Dalradian metamorphism in the Central Highlands of Scotland. Scott. J. Geol. 12, 236–68.

Roberts, D. (1981): Petrochemistry and palaeogeographic setting of the Ordovician volcanic rocks of Smøla, Central Norway. Norges geol. Unders. 359, 43–60.

Roberts, D. (1983): Devonian tectonic deformation in the Norwegian Caledonides and its regional perspectives. Norges geol. Unders. 380, 85–96.

Robins, B. and Gardner, P.M. (1975): The magmatic evolution of the Seiland Province and Caledonian plate boundaries in Northern Norway. Earth Planet. Sci. Lett. 26, 167–178.

Robinson, P. and Hall. L.M. (1980): Tectonic significance of southern New England. In: Wones, D.R. (ed.). Proceedings, The Caledonides in the U.S.A. Virginia Polytechnic Institute and State University, Mem. 2, 73–82.

Ryan, P.D. Williams, D.M. and Skevington, D. (1980): A revised interpretation of the Ordovician Stratigraphy of Sør Trondelag, and its implications for the evolution of the Scandinavian Caledonides. In: Wones, D.R. (edit.) The Caledonides in the USA, Virginia Polytechnic Inst. and State Univ., Mem. 2, 99–103.

Schenk, P.E. (1978): Synthesis of the Canadian Appalachians. Geol. Surv. Can. Pap. 78–13, 111–36.

Shackleton, R.M. (1954): The structural evolution of N. Wales. Lpool. Manchr. geol. J. 1, 261–97.

Siedlecka, A. and Siedlecki, S. (1967): Some new aspects of the geology of the Varanger Peninsula (Northern Norway), Preliminary report. Norges geol. Unders. 247, 288–306.

Siedlecka, A. and Siedlecki, S. (1971): Late Precambrian sedimentary rocks of the Tanafjord-Varangerfjord region of Varanger Peninsula, Northern Norway. Norges geol. Unders. 269, 246–294.

Simpson, A. (1968): The Caledonian history of the northeastern Irish Sea region and its relation to surrounding areas. Scott. J. Geol. 4, 135–63.

Smith, R.L., Stearn, J.E.F. and Piper, J.D.A. (1983): Palaeomagnetic studies of the Torridonian sediments, NW Scotland, Scott. J. Geol. 19, 29–46.

Smith, D.I. and Watson, J.V. (1983): Scale and timing of movements on the Great Glen fault, Scotland. Geology 11, 523–6.

Solyom, Z., Gorbatschev, R. and Johansson, I. (1979): The Ottfjäll Dolerites: geochemistry of the dyke swarm in relation to the geodynamics of the Caledonide orogen of Central Scandinavia. Sveriges geol. Unders. C756, 1–38.

Spjeldnaes, N. (1961): Ordovician climatic zones. Norsk geol. Tidssk. 41, 45–77.

Stephens, M.B. (1980): Occurrence, nature and tectonic significance of volcanic and high-level intrusive rocks within the Swedish Caledonides. In: The Caledonides in the USA. D.R. Wones (edit.). Virginia Polytechnic Inst. and State Univ., Mem. 2. 289–98.

Stephens, M.B. (1984): Evidence for Ordovician arc build up and arc splitting in the upper allochthon of Central Scandinavia. In: Gee, D.G. and Sturt, B.A. (eds.). The Caledonide Orogen – Scandinavia and related areas. Wiley.

Stewart, A.D. (1975): 'Torridonian' rocks of western Scotland. In: Harris, A.L. et al. (eds). A correlation of the Precambrian rocks of the British Isles. Geol. Soc. London, Spec. Rep. 6, 43–51.

Stillman, C.J., Downes, K. and Schiener, E.J. (1974): Caradocian volcanic activity in East and Southeast Ireland. Sci. Proc. R. Dubl. Soc. 5A, 87–98.

Storetvedt, K.M. (1973): A possible large-scale sinistral displacement along the Great Glen Fault in Scotland. Geol. Mag. 111, 23–30.

Strong, D.F. (1974): Plateau lavas and diabase dikes of northwestern Newfoundland. Geol. Mag. 111, 501–14.

Sturt, B.A., Pringle, I.R. and Ramsay, D.M. (1978): The Finnmarkian Phase of the Caledonian Orogeny. J. geol. Soc. London 135, 597–610.

Sturt, B.A. and Roberts, D. (1978): Caledonides of northernmost Norway (Finnmark). Geol. Surv. Can. Pap., 178–13, 17–23.

Sturt, B.A., Thon, A. and Furnes, H. (1980): The geology and preliminary geochemistry of the Karmøy ophiolite, S.W. Norway. In: Ophiolites, Proceedings of the International Ophiolite Symposium, Cyprus 1979, 538–554.

Swett, K. (1981): Cambro-Ordovician strata in NyFriesland, Spitsbergen and their palaeotectonic significance. Geol. Mag. 118, 225–250.

Thirlwall, M.F. (1981): Implications for Caledonian plate tectonic models of chemical data from the volcanic rocks of the British Old Red Sandstone. J. geol. Soc. London 138, 123–38.

Thorpe, R.S. (1979): Late Precambrian igneous activity in S. Britain. In: Harris, A.L. et al. (eds.). The Caledonides of the British Isles – reviewed. Geol. Soc. London, Spec. Publ. 10, 579–84.

Trettin, H.P. (1973): Early Palaeozoic evolution of northern parts of Canadian Arctic Archipelago. In: Pitcher, M.G. (ed.), Arctic Geology. Amer. Ass. Petrol. Geol. Mem. 19, 57–75.

Upton, B.J.G., Aspen, P. and Chapman, N.A. (1983): The Upper Mantle and deep crust beneath the British Isles: evidence from inclusions in volcanic rocks. J. geol. Soc. London 140, 105–22.

Van der Voo, R., French, A.N. and French, R.B. (1979): A paleomagnetic pole position from the folded Upper Devonian Catskill redbeds, and its tectonic implications. Geology 7, 345–8.

Vogt, T. (1967): Fjellkjedestudier i den ostlige del af Troms. Norges geol. Unders. 248, 1–59.

Walton, E.K. (1963): Sedimentation and structure in the Southern Uplands. In: Stewart, F.H. and Johnson, M.R.W. (eds.). The British Caledonides. 71–79. Oliver and Boyd, Edinburgh.

Westoll, T.S. (1945): A new cephalaspid fish from the Downtonian of Scotland, with notes on the structure and classification of ostracoderms. Trans. R. Soc. Edinburgh 61, 341–57.

Williams, H. (1978): Tectonic lithofacies map of the Appalachian orogen. Memorial University of Newfoundland Map No. 1, Scale 1:1.000.000.

Williams, H. (1979): Appalachian orogen of Canada. Can. J. Earth Sci. 16, 792–807.

Williams, H. and Hatcher, R.D. (1982): Suspect terranes and accretionary history of the Appalachian orogen. Geology 10, 530–6.

Williams, H., Kennedy, M.J. and Neale, E.R.W. (1974): The northeastward termination of the Appalachian orogen. In: Nairn, A.E.M. and Stehli, F.G. (eds.). The Ocean Basins and Margins, Vol. 2. Plenum, New York, 79–123.

Williams, H. and King. A.F. (1976): Southern Avalon, Newfoundland: Trepassey Map area. Geol. Surv. Canada. Pap. 75–1A, 11–5.

Williams, H. and Max, M.D. (1979): Zonal subdivision and regional correlation in the Appalachian-Caledonian orogen. In: Wones, D.R. (ed.) Proceedings, The Caledonides in the U.S.A., Virginia Polytechnic Institute and State University. Mem. 2, 57–62.

Wilson, J.T. (1966): Did the Atlantic close and then re-open? Nature, London 211, 676–81.

Winchester, J.A. (1974): The zonal pattern of regional metamorphism in the Scottish Caledonides. J. geol. Soc. London 130, 509–24.

Windley, B.F. (1983): Metamorphism and tectonics of the Himalaya. J. geol. Soc. London 140, 849–66.

Woodcock, N.H. (in press). Lower Palaeozoic sedimentation and tectonics in Wales. Proc. Geol. Ass. London.

Book Reviews

The Ocean Floor, Bruce Heezen Commemorative Volume

Edited by R. A. Scrutton and M. Talwani, 1982, John Wiley & Sons, New York, 318 pp, £ 47,50 (Hardbound), ISBN 0471100919

The encompassing title of the book is misleading. The scholar wanting to know everything about the ocean floor finds no balanced, comprehensive reference volume like "The Sea" and no subdivided introductory text like the contemporaneous 1982 publication "The Sea Floor". Instead, the reader is confronted with an assortment of different topics, supposedly reflecting the broad interests of the late Bruce Heezen. Indeed, the authors have not been constrained in their choice of subject matters, which is an easy approch for the editors to take but a tough one for the fair reviewer who searches for specific strengths and weaknesses of the publication.

The complete bibliography of Bruce Heezen (B. H.) compiled by M. Tharp, focusses on Heezen's lifelong interest in deep-sea sedimentation. From simple tools operated from the ocean surface, B. H. had advanced to direct observation when he found his tragic death on 21 June 1977, while working on the Reyjanes Ridge in the U.S. Navy's nuclear powered submarine NR-1. At the beginning of his career it also was the military that in 1952 classified all ocean depth contours and led Heezen to sketch his famous physiographic diagrams interpreted by an artistic method which was developed by professor Armin Lobeck of Columbia University.

Reminiscing on 30 years of collaboration with B. H. Marie Tharp describes the preparation of the enclosed World Ocean Floor Map. This was done by extrapolation and interpolation of trends observed in a succession of profiles. Compared to the sober arrays of SEA-BEAM plots B. H.'s physiographic maps resemble medieval cartography which reflects the personal touch of the master craftsman. Modern efforts in establishing ocean bathymetry make use of computers: the opening of the digital data bank of the US Navy World Scientific Ocean Floor Relief Model to the interested public is a recommendable policy, which warrants the quotation of the reference address: Dr. Leonhard Johnson, Code 461, Office of Naval Research, 800 North Quincy Street, Arlington, VA 211117.

Another paper prepared under US Navy auspices, by Ruddiman, Glover and Molfino, uses factor analysis of 13 parameters which measure the shape of mixed ash concentration vs. depth curves in 28 cores from the subpolar North Atlantic. Given the mind-boggling statistical tables, it is astonishing that the analysis can define the thickness of a fraction of a millimetre and the level of the initial ash deposition, which is later smeared by bioturbation and redeposition by currents. As the volcanic ash curves represent the sum total of all mixing processes (vertically and laterally), the paper may perhaps lead to a more precise definition of paleomagnetic stratigraphy.

Among the following articles which vary strongly in their degree of reference between their results and B. H.'s work the Russian papers are most commemorative. Udintsev and Kosminskaya in their cautious approach to sea-floor spreading concepts stress B. H.'s idea of the tectonic heterogeneity of the ocean floor, and offer their view that regional subsidence of continental crust to produce young ocean basins has taken place in the North Atlantic. This article, as well as a later one by Udintsev and Koreneva about the origin of the Ninetyeast and Broken Ridges in the Indian Ocean, has insufficiently labelled figures and no original Russian data. Thus they can only serve to refer the reader to the quoted Russian-language references.

Houbolt's paper demonstrates transport of sand along the length of shallow marine tidal sand ridges until at the end of the ridge, it becomes ready for a new depositional cycle. Thus the ridges move perpendicular to their long axes and to the tidal currents.

Uchupi et al.'s paper is a good documentation of erosional surfaces on the continental margin of NE America without proof of subaerial or submarine origin. Malahoff et al.'s paper on the sediment dynamics of Norfolk and Washington Canyons continues research in one of B. H.'s pet subjects, and describes the importance of bio-erosion for slope instability at crab 'excavation sites'. Such phenomena are best studied by submarines or by long term time-lapse photography. The nonspecialists, the changes of the ocean floor recorded in Thorndike et al.'s pictures may not appear drastic but one can imagine their effects after geological time spans. As human artifacts (cans, oil drums etc.) seemed concentrated in the canyon valleys of the US coast, this garbage-drain effect is supported by indirect evidence from the Mediterranean. Here, Le Pichon and Renard call for longitudinal transport by submarine avalanches that move cyclopean blocks and keep the narrowest canyon-portion free. They also interpret most of the erosional features in the thalwegs as deep-water phenomena of Quaternary age and therefore prefer a deep Messinian West Mediterranean Basin.

A morphological and geochemical study of hydrogeneous Mn-coatings by Nesteroff advises to look for waters enriched in dissolved metals and silica, if one is interested in accelerated Mn-nodule formation.

In their 1979 manuscript on NW African sediment dynamics, Seibold and Fütterer gave an overview of the activities of the Kiel marine geological group prior to the 1982 compilation of all German research on this margin by Springer Publishing House. With respect to the Quaternary period sedimentation, the Kiel school of actuo-geological concepts of marine geology admits that sometimes only the past may be the key to the past. In describing the paleogeographic and geologic setting for early man in Java, Nikovich et al. probably overemphasize the idea that sea level changes controlled 'Homo dubius' migration across the archipelago.

Restating his previously published ideas about the fracturing of Gondwanaland, Fairbridge challenges the geophysicists to search for a Late Paleozoic Argentina intracratonic-rifted-gulf extended between Namibia and southern Brazil. Vanney and Johnson prove in their marine geomorphology of the Kerguelen-Antarctica Passage that key geographic areas exist where it is particularly profitable to study the interaction of geodynamic and hydrodynamic phenomena. With a good background in the history of East Antarctic research they introduce many new geographic names, do away with others that do not exist (B. H.'s Gaussberg abyssal plain) and compile a physiographic map densely packed with 31 (!) different features. In the warmer realms of the Pacific, Menard ponders over the balance between rainfall and evaporation while researching an idea of Davis (1928) that the depths of lagoons and the morphology of island rings around atolls are primarily a result of solution weathering.

As a structurally oriented geophysicist with an interdisciplinary approach, the reviewer could best relate to two papers which combine results of different disciplines into one emerging picture. The paper by Corliss and Hollister is a synthesis of work done between 1976 and 1980 on a giant piston-core of the US Long Coring Facility obtained in the framework of the Seabed Disposal Program to consider the Central Pacific as a possible repository medium for high-level nuclear-waste disposal. In the analysis of the 24.5 m long core representing nearly 70 million years of continuous sedimentation a shocking truth for seismic interpreters emerges: the sound speed data are not obviously correlatable to the other variables studied. The construction of a stratigraphy based on (triangular) fish debris could become a field for English majors who must forget their geometry: no less than 86 triangle-types ichtyoliths are differentiated. Nevertheless, a paleoenvironmental model emerges that corroborates the NNW movement of the Pacific Plate through different sedimentation provinces, but remains unspecific about the dumping of nuclear waste.

The paper by Jones and Mgbatogu seems to be the centerpiece of the book (37 pp, 3 plates in evelope). In describing the structure and evolution of the Guinea Plateau, the authors find a part of the African margin which has not been extended by sedimentation since the early stages of continental separation from Florida. For this reason the continent/ocean boundary can be well defined. However, the seismic stratigraphy remains uncertain as no direct tie-lines to well control were used. The existence of Jurassic sediments between the Guinea and Sierra Leone Fracture Zones is also still questionable: the neglect of diurnals at such low latitudes could have impaired the identification of the magnetic anomalies. The hint from magnetic data and canyon structural control that the Guinea F. Z. at its eastern and appears to have been recently active (in Jurassic crust) could address a fundamental problem otherwise lacking in the book.

The book also lacks some of B. H.'s intuition of assembling various facts into fascinating hypotheses or pointing out controversies for further research. It can therefore not serve as a stimulating textbook for freshmen, but a graduate student planning a career in marine science can grasp the wide variety of scientific problems in this field.

Gerd Wissmann
Bundesanstalt für Geowissenschaften
und Rohstoffe
Postfach 51 01 53
D-3000 Hannover 51

Cyclic and Event Stratification

By G. Einsele and A. Seilacher (Eds.). Springer-Verlag, Berlin/Heidelberg, 1982, XIV + 536pp, 180 illustrations.

This book results from a successful Rundgesprach held in Tubingen in April 1980 when some 43 sedimentologists gathered to consider one of the most basic features of sedimentary rocks — the problem of bedding. The majority of these experts came from the host institution, the Institut fur Geologie und Palaontologie Universitat Tubingen, with seven from U.S.A., nine from F.R.G., five from U.K., one from Switzerland and one from the Netherlands. In the preface the editors state "The title of the symposium focuses on the key questions: to what extent does bedding reflect the gradual cyclic and periodic changes of our telluric system or rather rare and unpredictable events that occur in almost any sedimentary regime?" They go to highlight the inherent dilemma that commonly splits researchers into cyclist and catastrophist campls, those who record sedimentary cycles and those who record event deposits, both views are presented in the volume. This is a strictly academic topic and most geologists have satisfactorily logged successions and correlated sequences without enquiring to this depth about ultimate origins. The Tubingen Institut, however, is clearly far ahead in this field of sedimentology and they were able to set the problems and seek interpretation from their invited speakers.

In the book the editors have divided the 44 papers (all in English) presented at the symposium into three sections. Part one, consisting of ten papers, is concerned with cyclic sequences or periodites: part two, of 22 papers, concerns event stratification; a rather shorter part three of eleven papers, five of which are brief abstracts, deals with cyclicity and event stratification in black shales. Thus event stratification dominates the book with 267 pages in part two and additional contributions in part three. Cyclicity is coverd in 158 pages in part one and some additional coverage in part three. The book ends with a short summary of five pages by the editors which provides a 'simplified overview' of the occurrence and palaeogeographical significance of event deposits and truly cyclic sediments.

Cyclic deposits or periodites are introduces by G. Einsele in the first paper of section one. He concludes that they are commonly obliterated or obscured by secondary processes; they are difficult to recognise if not enhanced by diagenesis; they are easily overlooked if they alternate with turbidites. Good preservation and recognition of periodites appears to be restricted to marl-limestone sequences and all the papers about periodites in part one of the volume deal with these successions. The same author next provides a long paper on the diagnosis, significance and causes of periodites. This is a major contribution and backed up by many illustrations and a comprehensive bilbiography. Further papers deal with Mesozoic and Lower Carboniferous limestone-marl sequences and bring in the importance of modification owing to diagenesis; M. Walther describes a limestone-shale alternation from NW Ireland that he shows to be largely diagenetic in origin. The mechanism of formation of truly cyclic sediments is tackled by W. H. Berger using data obtained from deep sea cores (Glomar Challenger – D.S.D.P.) that suggest that the evolution of deep sea environments is not gradual, but punctuated by periods of rapid change. While W. Schwarzacher and A. G. Fischer deal with cyclicity related to perturbations of the earth's oribt at 100 000 year intervals.

Event deposits are introduced by A. Seilacher in a short paper that comes at the beginning of part two. He shows that rare, but geologically common events such as severe storms, floods, turbidy currents, seismic shocks and volcanic eruptions leave behind stratigraphical units with distinctive sedimentological and ecological features and successions. The potential application of event stratigraphy in correlation, basin analysis and evolutionary research is emphasised. In a later paper the same author describes the distinctive features of sandy tempestites and shows that sedimentological and biological features allow the distinction of wave-generated storm sands (tempestites), flood deposits (inundites) and current-generated sandy deposits (turbidites). Other authors in part two of the book deal with storm dominated stratifaction in the Muschelkalk of Germany, Eocene of Egypt, Cretaceous of southern France, Lower Lias of south Germany and late Precambrian of southern Norway. Methods of interpretation are also described with R. Schmidt dealing with belemnites as current indicators in turbidites, R. Goldring and T. Aigner discussing the significance of scour and fill and D. I. Gray and M. J. Benton showing that multidirectional palaeoccurents are an indicator of shelf storm beds. There are also unexpected topics such as J. Winter on the crystal forms of zircon related to the depositional history of volcanic sediments. These papers are all well illustrated, maps, sections, tables and photographic plates are well chosen and drafted, reproduction on the printed page is excellent.

In part three the cyclicity of black chales is introduced by A. Wetzel in a major paper concerned with a hierarchy of cyclic events, mega-, macro-, meso- and micro-scale cyclicity, representing periods of the order of one hundred million years down to between one year to one hundred years. The organic content of black shales is discussed by several authors dealing with Triassic, Jurassic and Cretaceous.

Separation of truly cyclic from event intercalations is described by J. Paul in the black shales of the Kupferschifer who finds three grades of stratification: alternating varvelike layers of probable seasonal origin, cycles of carbonate rich and shaly layers caused by fluctuation in plankton productivity and shell-bed tempestites. J. P. Walzebuck finds evidence for similar cyclic and event stratification in the Toarcian black shales of NW Greece.

From the commentary on the three sections of the book it will be clear that this is an authoritive research compilation written by the informed expert for professional readership. It is in no way a textbook and little space has been found to introduce the subject matter of the book to the reader. To make matters more difficult for the general reader the subject is new in sedimentology and with it there has grown a new terminology. Words such as — periodites, rhythmites, laminites, tempestites, seismites, contourites, inundites, etc. are descriptive terms that presume an origin in many cases, clearly they should be used with caution and as they grade and alternate one with another, exact definitions are vital. Definitions are given in the text of the book, but to find them without a glossary or index is slow and frustrating, this is my only real criticism of the book. A definitive glossary of new and significant technical terms would have been a valuable addition to the work, firstly to pin down the expert on the exact usage of the terminology and second to help the layman through this maze of jargon with which even most geologists are unfamiliar. The other omission is the lack of any form of index to the book. There is no way of finding a fragment of detail or commentary that the reader needs to see again. No doubt the production of an index would have delayed the time of publication and increased the price of the book, but as a research tool it would have greatly increased its usefulness and transformed it from a valuable collection of research papers into a fully fledged reference book. The publishers should bear in mind these comments with an eye to the future. Surely this book would have been a better commercial proposition had it been blessed with a definitive index.

These are minor criticisms of a remarkable book, the more so because it is what the editors call a "home-made product" produced in camera-ready copy with all its high quality tables and illustrations by the Tubingen University Institute for Geology and Palaeontology. The publishers, the Institute at Tubingen and especially the editors are to be warmly congratulated on their achievement in producing an excellent publication which will, no doubt, be a standard reference in sedimentology for many years to come. After studying this book one thought in particular emerges — what an interesting and progressive seminar it must have been in Tubingen in April 1980!

G. A. L. Johnson
Dept. of Geological Sciences,
University of Durham,
U.K.

Geophysical Fluid Dynamics

By J. Pedlosky. Springer-Verlag, New York, Heidelberg and Berlin, XII, 624 pages, 1979, reprinted with soft cover 1982

Fluid dynamical phenomena abound in the outer layers and interiors of planets and stars, including the Earth and Sun. The observational evidence takes very many forms, and only in the case of phenomena occurring in the Earth's atmosphere and oceans have detailed direct observations of flow velocity, pressure, etc., been made over useful periods of time. Studies of convection in the Earth's mantle for instance, are largely based on near-surface measurements of heat flow, gravity, etc, and studies of core motions on measurements of the surface geomagnetic field.

The term "geophysical fluid dynamics" was coined about a quarter of a century ago. It has no precise definition, but in its widest sense it is a useful label for the study of basic hydrodynamic (and magnetohydrodynamic) processess underlying fluid dynamical phenomena encountered in the study of planets and stars, as elucidated by theoretical investigations and related laboratory and numerical work. Geophysical fluid dynamics laboratories are now found in many institutions and the use of the prefix "geo" when objects other than the Earth are involved bothers only the pendants (cf. geometry).

Good books with misleading or ambiguous titles are not uncommon and this is one of them. But the title "Geophysical fluid dynamics" is certainly shorter and more voguish and eye-catching than "Some carefully-worked mathematical problems in the study of rotating stratified fluids for graduate students with an interest in large-scale motions in the Earth's atmosphere and oceans", which more accurately describes the main contents of this valuable contribution to the fluid-dynamics literature by a leading worker. The first chapter is a short one giving a good account of the standard basic dynamic and thermodynamic equations governing the flow of a rotating stratified fluid and introducing basic parameters such as the Rossby (or Kibel) number and the Rossby (or Prandtl) radius of deformation. This is followed by another short chapter presenting several standard theorems governing vorticity, circulation (Kelvin), potential vorticity (Ertel), and relationships governing geostrophic flow (Taylor — Proudman theorem, thermal wind equation). The rest of the book apart from a bibliography and index, comprises six long chapters averaging eighty pages in length. The first of these, chapter 3 entitled "Inviscid shallow-water theory", considers several types of small-amplitude wave motion, including the well-known Poincaré, Kelvin and Rossby waves, introduces Rossby's useful "beta-plane" for simplifying the equation governing motions on a sphere, and discusses resonant interactions associated with weak non-linearity. Dissipative effects enter for the first time in chapter 4 ("Friction and viscous flow"), which gives the theory of the Ekman boundary layer, including the important process of boundary-layer suction, and considers applications to "spin-down" and other processes in homogeneous fluids involving frictionally-induced changes in potential vorticity as well as the decay of Rossby waves and the structure of side-wall boundary layers when non-linear effects are negligible (Stewartson layers). The southward Sverdrup drift in simplest "beta-plane" models of the wind-driven circulation of the oceans (in which density variations are neglected) is associated with Ekman suction at the surface. The return flow occurs in a highly ageostrophic western boundary current reminiscent of the Gulf Stream in the Atlantic Ocean and the Kuroshio Current in the Pacific Ocean; chapter 5 presents a systematic treatment of the original theoretical model of this phenomenon, due to Stommel in 1948, and of many subsequent variations on that theme, including some numerical studies.

Chapter 6 entitled "Quasi-geostrophic motion of a stratified fluid on a sphere" treats a variety of problems bearing on dynamical meteorology and oceanography, including Rossby waves in a stratified fluid, forced stationary waves in the terrestrial atmosphere, theorems governing interactions between waves and zonal flow, and the structure of the ocean thermocline. "Instability theory" is considered in some detail in chapter 7, where the mathematical analysis of incipient baroclinic waves in continuous systems (Eady and Charney) and two-layer systems (Phillips) is given, and some effects due to friction and non-linearity are also discussed. Chapter 8, the final chapter, treats "Ageostrophic motion", including continental-shelf waves, theory of frontogenesis and waves in equatorial regions.

The publisher's claim that this book "offers a clear and logical contribution to understanding geophysical fluid dynamics" is, in this reviewer's opinion, justified. But only when it is read in conjunction with material that is more closely linked with observations and experiments can the book provide "students and scientists with the necessary background for understanding and pursuing research in oceanography and meteorology".

R. Hide
Geophysical Fluid Dynamics Lab.
London Road
Bracknell, Berks, RG12 2SZ
U.K.

Precambrian Plate Tectonics

Developments in Precambrian geology No. 4. Editor: Kröner, A. 1981, XXI, 781 pages, 177 figs. 26 tabs. Elsevier; Amsterdam. ISBN 0-444-41863-6

The 1960s saw the birth of the theory of plate tectonics, developed from observations of present movements of the lithosphere and the interpretation of the ocean basins. In the 1970s the theory was tested and confirmed and extended back in time to the period before the direct record of the oldest parts of the present ocean basins. Thus the theory was used to explain the development of Phanerozoic orogenic belts by the opening and subsequent closure of ocean basins in a Wilson Cycle and gave rise to the revolution in the Earth Sciences. Prior to the 1980s the relevance of plate tectonics to the majority of Earth's history — the Precambrian — had remained enigmatic. This volume takes a firm grasp of the issue and discusses exhaustively the relevance of almost every aspect of Precambrian geology to plate tectonics. Although no doubt discussion and argument will continue, this volume sets the scene for the 1980s being the decade when the theory of plate tectonics became modified to cover the Precambrian Era.

The volume consists of 28 individual contributions involving 40 authors from 14 countries. Although this should have resulted in a global coverage, most of the regional contributions describe the geology of either the African or North American Shields with only rudimentary coverage of the Australian, Indian, Baltic and Siberian Shields. This probably reflects a very real inbalance of our knowledge of the various Shields.

The contributions have been grouped into 7 sections each of which covers a distinct aspect of the topic. The first section contains four reviews of Precambrian evolution which inevitably build on generalised geological relations but also rely heavily on theoretical concepts, particularly on the exponential drop in heat flow during Earth's evolution. Each of the reviews agrees that a modification of modern plate tectonics is required to explain the Precambian developments and indeed this is a recurrent theme throughout the book. It is in this section that the Prearchaean period (4.5—3.8 Ga) is most widely treated but even here very little attention is paid to it and I would have liked to have seen a distinct section discussing this interesting and contraversial period of Earth history.

The second section deals with Archaean (3.8—2.5 Ga) tectonics and contains 5 contributions. Although the various granulite-greenstone belts described are superficially similar, they show marked differences in detail which give rise to differing conclusions about the nature of the early Archaean crust and the variant of plate model required. They are all agreed, however, that observations tend to support the existence of a high geothermal gradient producing thinner and more active plates. Perhaps one of the most useful aspects of this and the next two sections is the compact and clear descriptions of particular Precambrian terranes.

In section 3 four contributions describe Lower and Middle Proterozoic (2.5—1.2 Ga) tectonics. It is unfortunate that all four deal with the North American shield and consequently one is left uncertain about the applicability of the discussion to other shields. This is particularly so since different models are produced for all four regions described; ranging from an essentially modern plate tectonic analogue for the evolution of the Mid Continent region of North America (Van Schmus & Bickford), through a curious ring model of oceanic opening and closure (incipient Wilson Cycle) for the circum-Superior Province belt (Baragar & Scoates) to en-sialic subduction and collision models for the Labrador geosyncline (Dimroth) and the Grenvillian orogeny (Baer).

The transition from Upper Proterozoic to Lower Palaeozoic (1.2–0.6 Ga) tectonics is excellently described in the three contributions on the Pan-African event (Section 4). In almost every aspect Pan-African plate tectonics are shown to be comparable with modern tectonics. In particular, destructive plate margins producing island arc volcanic activity, back-arc basins, obduction of ophiolites and collisional orogenesis, are almost indistinguishable from Phanerozoic examples.

The final three sections of the book describe and analyse specific aspects of Precambrian geology. In Section 5 geochemical and geophysical data are presented in four contributions and their implications for plate tectonics discussed. It is significant that in each case the authors argue for an evolution of lithospheric events such that the change to present-day tectonic subduction regimes took place during the Proterozoic. Precambrian palaeomagnetism is discussed in four contributions in Section 6. A large amount of new palaeomagnetic data is given for Laurentia, the Balto-Russian Shield and Gondwanaland but almost all these data are confined to the Proterozoic. Apparent Polar Wandering Paths show the now familiar hairpin bends that signify major relative movements of the continental shields and thus are not incompatible with Phanerozoic plate tectonics. In the final section two contributions discuss the relationship between Precambrian ore deposits and plate tectonics.

It should now have become clear from the above that this work contains a wealth of information and well informed discussion about a subject that has become paramount in many geologists' minds. The danger that such a volume contains an unconnected jumble of contributions has been skillfully overcome by painstaking and unobtrusive editing and cross-referencing. I found very few typographical errors and the standard of draughtmanship is generally very high. The contents list (11 pages) and index are excellent.

Although the essence of some of the contributions has already appeared in earlier publications, this is a book that certainly no Precambrian geologist can be without and many geologists working in the Phanerozoic will also want to own a copy. In my opinion it will become one of the classics of the geological literature.

R. A. Gayer,
Department of Geology,
University College,
P.O. Box 78,
CARDIFF CF1 1XL, U.K.

Mineral Deposits and the Evolution of the Biosphere

H. D. Holland & M. Schidlowski (eds.), Springer Verlag, Berlin, F.R.G., 1982, 332 p., $ 19.00 (U.S.)

This book is a collection of articles and state-of-the-art reports presented as background of the Dahlem Workshop on Biospheric Evolution and Precambrian Metallogeny which was held in West-Berlin, between September 1 and 5, 1980. Its goal, as stated in the cover page, was "to define evolutionary relationships among biological processes, the ocean-atmosphere system, and the formation of sedimentary mineral deposits". This is a more accurate description of the contents than the title. Mineral deposits are specific topics of only 3 of the 12 articles and of 2 of the 4 State of the Art reports. Actually, the biology and biochemistry of early life forms, their impact on the chemistry of the biosphere, and the formation of sedimentary mineral deposits receive approximately equal treatment in this very interesting volume.

The general format of the conference, which was to bring together specialists from various branches of science into a single interdisciplinary workshop, is reflected in the book which will best be appreciated by scientists acquainted with the problems. The topics are grouped into three sections: "microbial processes and ecosystems" deals mainly with biochemistry, "the morphological and chemical record of the Precambrian biosphere" mainly with paleobiology, and "biological processes and the formation of mineral deposits" with problems of economic geology and geochemistry. The results of discussions of the four interdisciplinary groups of participants are presented in the state-of-the-art reports at the end of the volume: 1-Sedimentary iron deposits, evaporites, and phosphorites; 2-Stratified sulfide deposits; 3-Reduced carbon in sediments; and 4-biochemical evolution of the ocean/atmosphere system.

The first part of the book is devoted to the biochemistry of microbial systems involved in the cycles of sulfur, carbon, and iron with an attempt to retrace the biochemical evolution of some geologically significant microbial communities. The subjects are discussed in three papers by H. G. Trüper, K. L. H. Edmunds et al., and K. H. Nealson. To the non-initiated this biochemical discussion shows how intricate and diverse microbial processes are and serves as a warning against oversimplifying hypotheses. Three interesting conclusions stand out among many others. The photosynthetic oxygen production may have been preceded by an accumulation of sulfate in sea water which ceased when assimilatory sulfate reduction was invented some 2.8–3.1 Gy ago (H. G. Trüper); chelated compounds may have been important agents of Fe transport (K. H. Nealson); and lipids may serve as evolutionary indicators of micro organisms (K. L. H. Edmunds, et al.). The technical language may be, however, an obstacle to non-biochemists. Interested but not fully initiated readers should expect to peruse this section several times to extract all its informative details.

The second part of the volume offers the latest account of the Precambrian biologic record and the techniques which are being used to track it down. Curiously, it is the last paper of this part which deals with the very beginning of life, the prebiotic synthesis of organic compounds (S. L. Miller). The fossil evidence is presented by S. M. Awramik, J. Langridge provides a geneticist's viewpoint, and the chemical residues of life are described by M. Schidlowski for carbon and by D. M. McKirdy and J. H. Hahn for kerogen and hydrocarbons. Of the many fascinating conclusions a few are that life developed rather rapidly in somewhat cool sea water as pre-organisms formed from an atmosphere of CH_4, H_2O, N_2, NH_3, or perhaps CO_2 with H_2. It later evolved into organisms which developed oxygen producing photosynthesis which may have left its imprint in the 3.8 Gy old Isua metasediments of Greenland. The antiquity of life on earth is also partially substantiated by studies of carbon isotope ratios and analysis of organic matter, but the specialists insist on the difficulties in building well founded hypotheses on such fragile remains. This second part of the book stands out by its excellent balance between quality of information, clarity of style, and stimulating presentation of future research goals.

The third section focuses on the formation of sedimentary mineral deposits and sediments. Banded iron formations are treated expertly by H. L. James and A. F. Trendall, who deserve special praise for the rigor of the analysis and the perceptiveness of their geologic interpretations (e.g. the distinctions between syngenetic and diagenetic banding in the BIFs). They make a solid case for a deposition of the BIFs as a result of seasonal evaporation on shallow open shelves near arid landmasses. In a very different vein A. Lerman's assessment of the sedimentary mass balance through geologic time is also a remarkable investigation which applies quantitative geochemical reasoning to interrelations of the cycles of carbon, iron, phosphorus, and calcium. Some of his verifiable inferences are likely to spur future inquiries. On the other hand the discussion on the origin of stratiform sulfide deposits (P. A. Trudinger & N. Williams) is somewhat uneven. It is true that the topic's complexity and its polemic nature make it difficult to give a treatment which would win unanimous approval. Their case for a lack of evidence for biogenic sulfide deposition in modern environments is well documented, but their reliance on sulfur isotope data for extending the view to Precambrian deposits is later made questionable in the second State of the Art Report. Also, their statement that "large scale sulfide deposition, both ancient and modern, appears to be restricted to environments with clear-cut hydrothermal associations" is likely to dissatisfy many non-English speaking European geologists, especially because few of their works are listed in the references. Perhaps the authors' appeal to establishing criteria for the distinction between syngenetic, epigenetic, and diagenetic criteria may show eventually that organisms play a larger role during diagenesis than deposition. The paper by R. E. Folinsbee dealing with the variation of the distribution of mineral deposits with time is more of an overview of a very vast area of study than a detailed investigation. It offers the only description of the Archean and Proterozoic Au, U deposits (Rand/Blind River type), but it is short. One would have expected that these deposits would be devoted a larger place in such a conference.

The four state-of-the-art reports which synthesize the discussions of the interdisciplinary groups represent a very laudable idea which seldom is found in conference proceedings. They all conform to a similar outline: a synthesis of the most acceptable conclusions is followed by a list of outstanding problems and questions to be addressed to in future inquiries. The report on Reduced Carbon in Sediments (J. H. Oehler, rapp.) offers an excellent balance between presentation of conclusions and future goals. The group's well substantiated conclusion that there is no compelling evidence of a causal link between microorganisms and banded iron formations is likely to have a profound impact on future hypotheses. The group on the Biochemical Evolution of the Ocean-Atmosphere System offers specific constraints regarding the time of appearance of life and the composition of the atmosphere and oceans which allowed this event. On the other hand, the group is forced to admit that many specific details continue to elude them: the time of appearance of O_2 producing photosyntesizers, the O_2 production rate, Archean and Proterozoic weathering processes, etc.

The report of the Stratified Sulfide Deposits Group (N. Williams, rapp.) distinguishes clearly the many outstanding problems into metal sources (most likely from rock water interaction) and sulfur reduction mechanisms during deposition and diagenesis. In general, the group's conclusion are that Precambrian stratified sulfide deposits offer little potential as source of information for biogenic processes. The Group on Sedimentary Iron Deposits, Evaporites and Phosphorites (A. Button, rapp.) seems to have centered its efforts to unravel the problems of the BIFs. The relationships between iron deposition and photosynthesis are explored in great detail, and the various options are presented with great objectivity. In contrast, the sections on evaporites and phosphorites are short and mainly designed to show the vast amount of information which remains to be collected.

In summary, Mineral Deposits and the Evolution of the Biosphere is an invaluable source of information for anyone who wishes to become acquainted with the latest advances in the fields of Precambrian paleobiology and sedimentary petrology. The volume belongs in any specialized library to serve as reference or as guide for future research. The reasonable price of the book will he appreciated by those who wish to add it to their personal collection.

Alexis Moiseyev
Department of Geological Sciences
California State University
Hayward, CA 94542
U.S.A.

Science at Sea: Tales of an Old Ocean

W. H. Freeman, San Francisco, 1981, 186 pages, By Tjeerd Van Andel.

This is a delightful book that is now appearing in a second edition. Having a first edition that appeared as recently as 1977 it must have done reasonably well and the reviewer can say only that he is pleased. The enterprise must have been something of an act of faith on the part of both author and publisher.

The author is a polymath with a silver pen. Writing for the adult lay reader he unfolds the fascinating history of the present day ocean basins and how that history itself came to light during the explosion of new ideas and the startling discoveries of the last two decades.

Perhaps the point that should be made most emphatically is that the book is *not* a text. It is not, and does not pretend to be, a rigorous and systematic treatment of the subject. Rather it is an idiosyncratic and charming ramble along the pathways of ocean science, exploring its many facets and with many a sideways glance at its bearing on a variety of contemporary issues of importance.

The author has interspersed a main career as a professional marine geologist with periodic excursions into professional archaeology, supported by an abiding interest in that subject. This gives his writing an unusual width of perspective and contributes much to its charm. After an early pair of chapters in which he introduces his readers to continental drift, sea-floor spreading and plate tectonics, Van Andel shows how the patterns of ocean water circulation must have changed with time and how this has been a major control on the distribution of life forms. The oceans play a major role in controlling the Earth's climatic pattern and as the oceans have changed in size and shape, the climate has responded.

One expression of the climatic changes, but not solely attributable to changes in ocean circulation, are the ice-ages. In turn they withdrew great volumes of water from the oceans, trapping this water in polar ice caps, lowering sea level and allowing the continents to emerge. But changes in sea level have other causes too ... and so the story continues.

Van Andel is not writing for students or professionals in the field, but very few of them would read the book without gaining some new insight.

Some would be irritated by some of the author's physical explanations where he has been obliged to oversimplify in the interests of intelligibility. Most, however, will be mollified by his intriguing account of the world trade in salt in ancient times.

This seems to the reviewer to be an excellent book for the interested layman, but one that will be read and enjoyed by many who deem themselves knowledgeable. It should be on the shelves of Geology libraries.

E. R. Oxburgh
Department of Earth Sciences
University of Cambridge
U.K.

The Fossil Record and Evolution

Readings from *Scientific American* with Introductions by L. F. Laporte. W. H. Freeman and Co., 1982, 225 pp., £ 8.40 (softback). ISBN 0-7167-1402-7.

Because of the high publication standards of the parent journal, the thematic collections of *Scientific American* articles that are published from time to time always make interesting reading. This collection comprises sixteen articles, grouped into five sections, with a preface and with an introduction to each section provided by Léo Laporte. The theme for this collection is the fossil record of the history and evolution of life and the book is primarily aimed at students taking elementary palaeontology and evolutionary biology courses.

The interest and quality of the articles is uniformly high, but together they cannot honestly be said to form a particularly coherent or comprehensive text. Rather, the section headings (I The phenomenon of evolution; II Earliest traces of life; III Interpreting fossils; IV Some major patterns in the history of life; V Human evolution) should perhaps best be viewed as convenient loose categories for sorting out a clutch of articles in the general subject area that happen to have appeared in the pages of the parent journal. This restricted pedigree also means that, although 'virtually all the articles are popular classics written by the very people who made many of the original contributions to evolutionary theory and the history of life' (Preface), they were not themselves seminal works; they were more in the nature of popular summary articles written after their contributors had made their mark.

Five of the articles come from the splendid September 1978 issue of *Scientific American*, on 'Evolution' (though Valentine's article on 'The Evolution of Multicellular Plants and Animal's is incorrectly dated December 1978). These articles are as valid and useful today as when they were first published, barring a few minor points, such as, for example, the oldest stromatolites now known (from Australia) being older than the 3 billion years shown at the end of Schopf's article on 'The Evolution of the Earliest Cells'. And the same can be said of the two articles from later issues. However, some of the older articles (the oldest being Glaessner's 'Pre-Cambrian Animals' from March 1961) are beginning to show their age and I would hesitate to recommend them to a student without reference to some later works as well. For example, Newell's 'The Evolution of Reefs' (June 1972), though an excellent article in its time, dates from when it was still fashionable to emphasize the similarities between fossil and Recent tropical reefs ('the oldest ecosystem in earth history'), so much so that one might imagine that a virtually identical ecosystem bounced back after each extinction episode, with only the taxonomic affinities of the constituent organisms changed. During the course of the '70's, perhaps particularly in keeping with the range of possibilities (walled-reef and knoll-reef complexes and down-slope mudmounds) described by J. L. Wilson (1974), greater emphasis came to be placed on some of the differences in character, such as hydrodynamic context, between various fossil and modern reefs or buildups.

On the other hand Seilacher's even older article on 'Fossil Behavior' (August 1967) still reads with great freshness and would still serve as a useful introduction to trace fossils.

Overall, then, the articles are certainly worth having, their only faults being due to the ravages of time. Unfortunately Laporte's supporting passages do not really match them in quality. They are neither profound nor particularly useful; they do little more than briefly summarise the articles — in some cases rather over-simplistically — and certainly fail to provide the sort of updating promised in the Preface. For example, to return to Newell's article on reefs, it simply isn't good enough in 1982 to say (p. 84): 'Although there have been many comings and goings of different kinds of organisms throughout the Phanerozoic, the basic organization of a "reef" has remained intact'. These passages seem to be pitched at a much lower level than the articles themselves, and frequently have too much of a 'gee-whizz' quality for my taste. For example, from the Preface: 'Each of us, in our own individual existence, bears for a fleeting instant the fire of life that has come directly down to us through countless generations over billions of years. Think of it. You and I, quite literally, trace our origins directly back ...' etc. etc. We know he can do better than that from his very handy little book on 'Ancient Environments' (Laporte 1968). I wonder if he really gave the task the attention it deserved?

Another shortcoming of the Introductions is their failure to 'rock the boat'. *Scientific American* articles commonly adopt a fairly orthodox tone; even Lewontin's article on 'Adaptation' is mild compared with his own utterances elsewhere (e.g. Lewontin 1983). I would therefore have hoped for some references in the Introductions to, say, arguments concerning the reasons for the explosive early Phanerozoic radiation of metazoans: was it essentially environmentally driven, as favoured by Valentine and Moores, in their article on Plate Tectonics and the History of Life' (April 1974), or was it rather a simple consequence of exponential diversification, not requiring an environmental motor, as explained by Sepkoski (1978). Again, there is no reference in the comments on Schopf's article to any debate about the character of the primitive atmosphere and the context for the early evolution of life (see, for example, the various views expressed in Ponnamperuma 1983).

These weaknesses of the supporting text mean that if the book is to be used as a 'reader' to accompany introductory palaeontology classes, then the lecturer involved will certainly need to append some comments and bibliographies of his or her own.

With these qualifications, I do recommend the book, certainly to libraries servicing palaeontology and evolutionary biology classes. And many lecturers will find it a useful compendium to have on their shelves. Whether or not students will want to pay the not exactly modest price for it depends on how much effort their lecturer will wish to devote to serving up its undoubted but poorly cultivated merits.

L. F. Laporte, 1968. *Ancient Environments*. Prentice Hall, 116 pp.
R. C. Lewontin, 1983. Gene, organism and environment. In: D. S. Bendall, ed., *Evolution from molecules to men.* Cambridge University Press, 273–285.

C. Ponnamperuma (Ed.), 1983. *Cosmochemistry and the Origin of Life.* D. Reidel Publishing Co., 386 pp.

J. J. Sepkoski Jr., 1978. A kinetic model of Phanerozoic taxonomic diversity. I. Analysis of Marine Orders. *Paleobiology* 4 (3), 223−251.

J. L. Wilson, 1974. Characteristics of Carbonate Platform Margins. *Am. Assoc. Petroleum Geologists. Bull.* 58, 810−824.

P. W. Skelton
Department of Earth Sciences
The Open University,
Walton Hall, Milton Keynes
Buckinghamshire
U. K.

Methods in Field Geology

By F. Mosely, W. H. Freeman & Co., Oxford and San Francisco, 1981. 211 p., 146 Figs. ISBN 0-7167-1294-6

"Fieldwork and geological survey form important parts of undergraduate courses in the geological sciences, but these topics are covered only cursorily in standard textbooks for geology students", the author states in the preface and explains that his book is an attempt to fill a serious gap in the literature.

It is evident that this book is based on the author's many years of experience in field geology. As the selected case histories (Part II) prove, he gained this experience in many different regions of the world which differ in geology, in the degree to which this geology is known, in climate, in morphology and in infra-structure. Geoscientists with a strong background in field work will develop − strongly influenced by their experience and their tasks − different techniques of work which may best be labelled "personal" approaches. This book is an account of a personal approach and this is one of its strengths.

The book is not exactly a "textbook" (at least not a "standard textbook") but rather an amply illustrated guide with explanatory text. Many of the illustrations are sketches which look as if they had just been copied from the field-book (and actually they have), demonstrating clearly the importance of sketches and graphs in fieldwork. This emphasis is another strong point of the book.

Part I (General Principles) centers on the use of aerial photographs in the preparation for, and the conduct of field work. These are highly important sections because the value of the use of aerial photographs in geological mapping is still often underemphasized (at least in countries like Germany where reliable topographic maps are easily available). Unfortunately, the photogeological procedures described and suggested in these chapters look rather complicated − actually such work can be done often much more easily and more economically. In addition the description of the geometry of aerial photographs is not exactly correct (ref. explanation to Fig. 6).

Part II (Case Histories) shows how an experienced field geologist approached different tasks and how he solved them. It is worth noting that Quaternary geology and geomorphological problems are also given due consideration. Above all it becomes evident that a field-geologist should use a multidisciplinary approach because "the extreme specialist in any one field cannot hope to solve the problems" (ref. p. 138).

In summary I read this personal account with much interest. This book can be recommended — not only to undergraduate students — but to all those who know the basics already and still like to learn how others approach and solve geological tasks.

Prof. Dr. Dietrich Helmcke
Institut für Angewandte Geologie
Freie Universität Berlin
Wichernstraße 16
D-1000 Berlin 33

Applied Geophysics for Geologists and Engineers

By H. D. Griffiths and R. F. King. Pergamon Press, Oxford 1981, 230 pages.

This book is the second edition of *Applied Geophysics for Engineers and Geologists* which appeared in 1965 and which was reprinted three times. The first edition was written mainly with the need of civil engineers in mind. It was mainly concerned with the applications of geophysical methods in site investigations. The scope of the original book was however already broadened to include the principles of geophysical methods used in prospecting for oil and other minerals, thus taking into account the needs of geologists, mining engineers and petroleum engineers.

Like the first edition, this second edition is mostly concerned with the fundamentals of geophysical applications. However, as the change of title of the book indicates, the emphasis has shifted in the second edition. Because the first edition was popular with students of geology, the chapters on the applications of geophysical methods in the search for ore minerals have been strengthened considerably.

The emphasis of the second edition remains on seismic and resistivity methods, but the sections dealing with electromagnetic and induced polarization techniques have been enlarged.

The authors also stress the fact that rewriting had become necessary because the subject had greatly advanced since they gathered together the material for the original version. One has to agree with the authors that geophysics in a rather short time has moved into a new era both because of the developments in electronics and in computer sciences.

Beginning with an introduction and a short description of the units to be used in geophysics, various methods are presented. Seismic methodology is described beginning with the physical background of seismic wave propagation and leading to the ray-path and travel-time presentation for simple models. A chapter on seismic surveys and their interpretation begins with seismic sources and recording systems and deals mainly with reflection and refraction surveys. The section on seismic reflection surveys shows the various steps of data processing which are necessary to obtain the final cross-section. A main part of the section on refraction interpretation is devoted to complications which may arise in the case of hidden or blind layers.

Electrical methods are introduced by outlining the theory of electrical resistivity surveying. A discussion on interpretation techniques is followed by examples of practical applications. Other electrical methods such as electromagnetic resistivity and induced polarization methods are shortly described.

The chapter on gravity is mainly concerned with gravity reductions, the use of simple models for interpretation, and finally practical examples of gravity applications. The principles of magnetic surveying are briefly presented, followed by some pages on borehole logging, radiometric surveys and remote sensing. The collection of problems at the end of the book introduces the reader into the practice of geophysical-data interpretation.

The authors have succeeded in describing succinctly the principles of the most essential geophysical methods and in presenting useful practical applications in various fields of mineral exploration and engineering. They point out, at the same time, some of the most essential difficulties and limitations involved in the application of geophysical methods. The book can be highly recommended to geologists and engineers who need insight into the applicability of geophysical methods.

A. Vogel
Institut für Geophysikalische Wissenschaften
Freie Universität Berlin
Rheinbabenallee 49
1000 Berlin 33

Numerical Dating in Stratigraphy

S. Odin (ed.), John Wiley & Sons, 1982, 1040p., £ 59.50

This enormous work appears in two volumes and boasts an authorship of 137 scientists from over 20 countries. Its principal aim is to examine the principles and to re-evaluate the data used to define the Phanerozoic time-scale.

Volume 1 (630 pages) contains 34 chapters, each by one or more contributors, and is divided into 6 sections. The first 4 sections deal with methodology. Section 1 is entitled "Methods of Correlation" and includes chapters on biochronology, geochemical events as a means of correlation, the strontium-isotopic composition of seawater as a geochronometer, and palaeomagnetic stratigraphy. Section 2, on "Isotopic Dating", contains chapters on physical decay constants, inter-laboratory dating standards, potassium-argon analysis (including the $^{40}Ar/^{39}Ar$ method), and fission-track dating. Section 3 is about "Utilisation of Sediments as Chronometers" and contains 9 chapters on isotopic dating of glauconites (referred to throughout this book as 'glaucony', plural 'glauconies') and clay minerals in sedimentary rocks, and the interpretation of the results in the light of much experimental work. Section 4 deals with "Utilisation of High-temperature Rocks as Chronometers" and has 3 chapters on the genesis of bentonites, the dating of bentonite beds, and the dating of plutonic events.

The last two sections in Volume 1 are concerned with calibrating the Phanerozoic time-scale, system-by-system, using what the authors regard as the best available geochronological data, with special emphasis on the European stratotypes. Section 5 has 7 chapters on various aspects of the Cambrian-to-Triassic, including discussion of new data and re-interpretation of older data, whilst section 6 does the same for the Jurassic-to-Palaeogene in 4 chapters. What may strike the reader as peculiar in scanning the chapter headings in section 5 is the use of the term "Caledonian times" to refer to the Cambrian-to-Silurian, and "Hercynian times" to the Devonian-to-Permian. Surely a gross misapplication of accepted geological terminolgy, if ever there was one!

Volume 2 consists of a lengthy, multi-authored introduction on the geological evidence for the most important stratigraphic boundaries in the Phenerozoic (including the Precambrian/Cambrian boundary), followed by 251 abstracts each of which consists of a brief, but detailed, presentation of a particular datum point by the stratigrapher and geochronologist who collected and measured it. Each abstract contains sections on stratigraphical, geochemical and analytical uncertainties, as well as a discussion on the relevance of the isotopic dates for the time-scale. Some abstracts also contain tables of analytical data and diagrams (e.g. isochrons, maps, etc.).

For ardent Phanerozoic time-keepers, Part 2 is perhaps the most useful part of the book, and will no doubt remain a standard reference for some time to come. However, it is not easy to use Part 2 because there seems to be no logic in the arrangement of the abstracts. They are all jumbled up together and it surprises me that they are not arranged in a logical, stratigraphical sequence.

Then follows a summary of the entire Phanerozoic time-scale by the editor and several co-authors, which includes their chosen best isotopic ages for the epoch and period boundaries, together with estimated uncertainties in these boundaries. There are some contentious issues here, not least in the lower Palaeozoic, where these authors draw the base of the Cambrian at 530 ± 10 m.y., the base of the Ordovician at $495 \pm ^{10}_{5}$ m.y., the base of the Silurian at

$418 \pm ^{5}_{10}$ m.y., and the base of the Devonian at $400 \pm ^{10}_{5}$ m.y. Much controversy currently prevails about the reduction in age of the base of the Cambrian by ~40 m.y., and about the reduction in the length of the Silurian by ~20 m.y., suggested by the present authors. I remain to be persuaded by this fairly radical revision of other proposed Phanerozoic time-scales of recent years. I believe that the evidence for the more conventional view is still compelling. But time will tell!

Part 2 concludes with a 43-page bibliography with more than 1500 entries, followed by 35 pages of author and subject indexes.

This is an immense treatise, parts of which will form essential reading to everyone concerned with the geological time-scale. The individual chapters in Part 1 are of extremely variable quality and there is a great deal of technical and experimental detail which seems rather remote from the principal topic of the book. True strength of purpose and moral character would be needed to read this book in detail from one end to the other, especially as some of the English is rather unidiomatic, without actually being funny. However, the "moment of closure of a glaucony" (page 297) conjures up a rather pathetic, if not exactly a scientific, image.

This book may be regarded as a tribute to the energy and enthusiasm of originator and editor Gilles Odin, who also appears as author or co-author of 17 chapters. However, I don't think that we have yet heard the last word on the Phanerozoic time-scale. This subject, like any other, would quickly get dull if we had all the final answers!

Stephen Moorbath
Department of Geology and Mineralogy,
University of Oxford,
Parks Road, Oxford OX1 3PR, U.K.

Regional Trends in the Geology of the Appalachian-Caledonian-Hercynian-Mauritanide Orogen

Paul E. Schenk (ed.), D. Reidel Publishing Comp., ISBN 90–277–1679–X 398 pp., Price: 140 Dutch guilders

This book represents a collection of articles that resulted from a NATO conference/workshop on the Appalachian-Caledonian-Hercynian-Mauritanide Orogen held in Fredericton, New Brunswick during August 1982. Many of the participants were members of Project 27 – "The Caledonide Orogen" – of the International Geological Correlation Programme and thus included experts on a range of aspects of Caledonide geology from the various countries through which the orogen passes. The two-day workshop, prior to the four-day plenary symposium, allowed discussion between members within five main geological groupings: Geophysics, Stratigraphy and Sedimentology, Volcanism and Plutonism, Metamorphism, and

Deformation, so that a common theme within each group could be followed.

The 39 contributions in the book are grouped under these five disciplines with individual articles concentrating on syntheses for a particular sector of the orogen. The sectors are broadly national boundaries and include the Appalachians in the USA; the Appalachians in Canada and Newfoundland; Ireland, Britain, Scandinavia, France and Morocco. A sixth category, headed „symposium" includes contributions that do not fit easily into the other groupings.

It is clear from the contributions that some of the five disciplines were far better organised than the others. In particular the sections dealing with metamorphism and deformation provide a collection of articles that give excellent summaries of regional developments. For example, each of the eight articles in the deformation section are based on standardised structural maps of the orogen showing the timing and intensity of deformation. For the purposes of these maps, five deformation events have been recognised, and contributors were asked to assess their regional developments in terms of these five events. Unfortunately the standard time-scale used does not conform to either the 1982 BP time scale of Harland et al. nor the 1983 Geological Society of America time scale. Notably the Cambro-Ordovician boundary is placed at 520 Ma instead of 505 Ma, thus making the stratigraphic age of the north Norwegian (Finnmarkian) orogeny and the north British (Grampian) orogeny somewhat problematical. In addition authors often have had to 'stretch' or 'squeeze' their deformation events to fit into the five-fold classification.

The seven papers in the metamorphic section are founded on regional metamorphic maps around which authors describe and discuss the particular developments in their sector. Some of these articles, particularly those by Fettes on "Metamorphism in the British Caledonides" and Long, Max and Yardley on the "Compilation Caledonian Metamorphic Map of Ireland" are excellent reviews of an extremely complex literature.

By contrast the sections on geophysics and stratigraphy and sedimentology consist of a muddled selection of articles with little in the way of a coherent theme.

The book has been produced by a camera-ready process to speed up publication and it is praiseworthy that the results of a workshop held in August 1982 should have become available by 1983. However, the editors have had to sacrifice accuracy and, to a certain extent, style for speed. This has produced some spectacular errors. For example we are told that: "The Iapetus Ocean is believed by many geologists to have been formed up of a crystalline continent during late Precambrian times" and "The Caledonian fold belt of western Scandinavia outcrops over a strike length of some 1800 km with a maximum traverse width of 30 km". A little more editorial scrutiny would have avoided including such odd phrases as: "relictual high pressure rocks". The relatively small format of the book has resulted in some extremely highly reduced maps for which one needs a magnifying glass to read the 0.5 mm high text.

However, taking all these production criticisms in hand, I find this book an extremely useful synthesis of the wealth of detail that has been collected on probably the most extensively studied of all orogenic belts. I would certainly recommend it to all researchers on the Appalachian-Caledonide orogen, who will find it an invaluable source of information.

R. A. Gayer
University College Cardiff
P.O. Box 78
Cardiff CF1 1XL
U.K.

Tidal Friction and the Earth's Rotation II.

P. Brosche & J. Sundermann (eds.), Springer Verlag, Berlin–Heidelberg–New York 1982, 344 p., 112 figs. ISBN 0-387-12011-4, US $ 28.00

This book comprises the proceedings of a workshop held at the Centre for Interdisciplinary Research of the University of Bielefeld during September/October 1981. This workshop was the second (in a series?) to be held under the title "Tidal friction and the Earth's Rotation" and the book has been published using an identical format to the first proceedings published in 1978.

The book, or more accurately the collection of papers, complements the earlier volume remarkably well. It both brings up to date and expands upon some of the earlier papers using the same authors while at the same time broadens the topic considerably by introducing additional research areas with additional authors. A good example of the former is the paper on the history of the "Earth's Rotation since 700 B.C." and good examples of the latter include papers on the "Earth's non-uniform rotation" and the "Gravitational heating of Jovian satellites by tidal friction".

The subject is truly multi-disciplinary with all the major disciplines of science (mathematics, physics, chemistry, geology and biology) involved to greater or lesser extent in the book. It thus requires a reader with a broad base in science to fully appreciate all the papers — alternatively there is something in the book for anyone who has a research interest fringing on this field! Each individual paper is generally of a very high standard which is an unusual occurrence nowadays in works of this kind. However many of the criticisms levelled at edited collections apply to this volume too, such as the lack of uniform coverage and the non-uniformity of symbols, units and abbreviations between papers. In fairness to the editors however it must be stated that they are aware of these problems and have done an excellent job overall in managing to publish the book so soon after the Workshop with very few errors.

The volume contains almost 50% more material than the first volume which indicates perhaps a renewed interest in this field in recent years since the quality is undiminished. The book is an additional and valuable up to date reference volume for both library and private collections, particularly if used in conjunction with "Volume 1".

R. J. Edge,
Institute of Oceanographic Sciences
Bidston Observatory
Birkenhead
Merseyside L43 7RA
U.K.

The Sea Floor

E. Seibold & W. H. Berger (eds.), Springer Verlag, Berlin–Heidelberg–New York 1982, 288 pp., 206 fig., 9 appendices. ISBN 3-540-11256-1.

The book is a very readable, introductory survey of topics in marine geology, resulting from the collaboration by two well-known marine geologists. The text is intended "for all who might be interested in the subject, including those without formal training in the natural sciences", but I am afraid that this statement reflects unwarranted optimism on the part of the authors with respect to American high-school science instruction. Maybe things are better elsewhere, but even so, terms like "endogenic forces" and "exogenic processes" (given on page 1 without explanation) are a bit difficult for the uninitiated.

The level of the book is somewhere between F.P. Shepard's "Geological Oceanography" and J. P. Kennett's "Marine Geology", but closer to Shepard than to Kennett. As such, it serves its original purpose very well, namely as a supplementary college text (the earlier, German version by E. S. is still used as such in Germany, as stated in the introduction). In addition, the book could serve well as a basic text in marine-geology classes for non-majors, and in all those situations where the amount of detail presented in Kennett's book is judged to be overwhelming.

"The Sea Floor" begins with a brief review of the history of marine geology. It then deals (in 2 chapters) with the origin and morphology of ocean basins, including a brief review of sea-floor spreading. Sediments and sediment composition are treated in the following chapter, followed in turn by a chapter on the effects of waves and currents. Next are chapters on sealevel changes, marine organisms, climatic imprints, and deep-sea sedimentation patterns. One could find fault with this particular arrangement of topics, but I suspect *no* arrangement of these topics would satisfy *all* users of the book. The final topics of the book are devoted to paleoceanography and ocean resources. Nine appendices deal with conversions between systems of measurements, and with basic geologic, chemical and biological information (e.g. geologic time scale, minerals and rocks, marine organisms). These appendices are useful aids when dealing with non-majors.

Treatment of oceanographic topics is somewhat uneven. For instance, there is good coverage of ocean life, but little information on oceanic volcanic rocks, reflecting the authors own research interests.

There are a few Germanisms in the book, almost unavoidable when proceeding from an original German text, but in no way detracting from the overall "flow" of the text. More disconcerting is the fact that I discovered at least one illustration still employing a "German"-type use of the decimal point. This is confusing.

The quality of the paper is reasonable, half-tones are good to excellent. In contrast, the quality of black-line reproductions is uneven although all of them are readable. The price, $21.50, is within the range of current textbook prices. Teachers of introductory earth-sciences classes will find this book well suited to their instructional needs.

D.A. Warnke
Department of Geological Sciences
California State University
Hayward, California 94542
U.S.A.

Report

Report on the 5th International Conference on Basement Tectonics, held in Giza, Egypt, October 16–18, 1983

The 5th International Conference on Basement Tectonics was held on the campus of Cairo University, Giza, Egypt, October 16–18, 1983. The conference was the latest in a series of such conferences which began in Salt Lake City, Utah in 1974 and which have since been held every two years in the United States or Europe. The 5th Conference was organized by the Faculty of Science, Cairo University; the Basement Tectonics Committee, Inc.; The Geological Survey of Egypt and, The General Petroleum Corporation. The Conference was sponsored by the U.S. Geological Survey, National Aeronautics and Space Administration and the International Union of Geological Sciences. The Conference theme was "Observing and understanding the deformation processes of the basement". Particular emphasis was given to basement processes and to economic applications of the results of basement studies. More than 150 scientists attended the Conference. The outstanding hospitality of the Egyptian hosts will be long remembered by most participants.

The influence of basement structure on the occurrence and evolution of geologic events in the overlying rocks is becoming increasingly appreciated by exploration geologists and geophysicists. The papers given at the 5th International Conference on Basement Tectonics reflect the progress that has been made over the past decade in the understanding of basement structure as well as the techniques of mapping and analysis. As with previous conferences, the emphasis of the technical papers was on the nature of basement rocks and basement structure, the controls exerted by the basement on the evolution of the tectonic environment and, on the location and character of minerals deposits of all types. Also emphasized were discussions of geophysical and geological techniques used in mapping and analysing basement structure and on the origin of lineaments and basement structural patterns.

Geographic coverage of the papers was extensive and included the United States, Canada, South America, Scandinavia, Europe and the Middle East, India, Australia and New Zealand. Just over 50% of the papers dealt with some aspect of Egypt and the Sudan, the Red Sea Rift and the Arabian-Nubian shield and, on this basis alone, judging from the detailed nature of many of the papers, the Proceedings volume for the Conference promises to be a valuable reference for those working in the complex geology in and around the eastern end of the Mediterranean.

It was obvious from the presentations and discussions that the concept of a primary basement control of the patterns and nature of post-basement geologic events and on minerals deposits was generally accepted and, also, that there was enough common experience among the participants to allow ready communication of ideas by the speakers and ready exchange of different points of view during discussions. The author can vouch from personal experience that such was not the case at the 1st Conference in 1974.

Even though there have been considerable advances in effective application of basements tectonics in minerals exploration, there remain many theoretical differences in points of view concerning origin and evolution of the basic structural features and structural patterns of the Earth's crust. Over the past decade, the presence of great lineaments as fundamental elements of the structural pattern of the Earth has become generally accepted. In the keynote address, however, Dr. Hodgson pointed out that other, perhaps equally important fundamental elements of the Earth's structural framework have been largely neglected with respect to both observation and analysis. These are the great ring structures of the Earth which are becoming increasingly recognized and which are being found to exist on the Earth to an extent similar to that of Mars, Mercury and the Moon. As fundamental Earth structures, however caused, they have influenced later structural and stratigraphical events within and along their boundaries. These features, along with large order lineaments properly comprise the fundamental structural framework of the Earth's crust. It will require intensive future investigation to fully understand the structural controls on later geologic events. Hodgson suggested that this should be a pressing matter for investigation over the coming decade.

A perennial area of disagreement among geologists is how to classify lineaments. Jules Friedman of the U.S. Geological Survey made a plea for resolution of this problem and informed the audience that a National Lineament Map for the United States is being compiled and agreement on a uniform scheme of classification is highly desireable.

The general high quality of the papers presented make it difficult to single out particular papers for comment. The relative impact of a given paper depended largely on the specific interests of the listener. Of considerable interest and representative of the kind of approach to basement studies in Egypt was the paper by El Hakim and El Saharty giving the relations between basement structure and mineral deposits as outlined by a regional magnetic survey in Egypt's Eastern Desert. Of interest to this author was the excellent review of the extensive mapping and analysis of photo linears in Egypt given by Hassan El Etr. Various other authors unravelled important aspects of the structure and tectonic history of the eastern Mediterranian region and North Africa. Interesting papers on tectonic provinces and selected large order lineaments of the United States were given by a number of authors including D.P. Gold, M.J. Aldrich, Yngvar Isachsen, S. Parker Gay, Jr., D.L. Baars, Patrick J. Barosh and others. Ivar Ramberg and Kjel Buer of Norway and Paavo Vuorela of Finland outlined applications of basement tectonics to waste disposal in non-sedimentary terranes. Kvet and Kopecky of Czechoslowakia outlined Landsat lineaments of Scandinavia. S. Bhattacharji and D. Jayakumar showed how basement controls have served to control mineralization in portions of the eastern and southern Indian Peninsula. B.L. Lammerer gave an important paper describing the fracture pattern of the northeastern Alps and how it is related to Alpine type deformation. H. McQuillan and Vadim Anfiloff outlined the role of basement structures in New Zealand and Australia. It is not possible here to review with justice all the excellent papers that were given and it can only be hoped that the Proceedings Volume for the Conference will be published as soon as possible and without the delays which plagued the Proceedings for the 2nd and 3rd Conferences.

The pre-Conference field trip lasted four days and was very well attended. It covered the general geology and tectonics of the basement rocks of the central part of the Eastern Desert of Egypt. Those who participated were impressed by the interesting and informative presentations of the trip leaders. It would be useful to have the field trip route and commentary included in the Proceedings volume.

Additional information concerning this Conference can be obtained from Dr. M. A. ElSharkawi (General Secretary), Department of Geology, Cairo University, Giza, Egypt or, Dr. John J. Gallagher, Occidental Oil & Gas Corp., 5000 Stockdale Highway, Bakersfield, CA, 93309, USA.

The 6th International Basement Tectonics Conference is scheduled to be held in Santa Fe, New Mexico, USA, September 15–20, 1985. A field trip is tentatively scheduled during the week of the Conference to Bandolier National Monument and Valles Caldera. A post-conference field trip is also planned but not yet scheduled. There will be special symposia on the basement of the Mid-continent of the United States, radiometric dating of Precambrian rocks and, exploration techniques and Precambrian Geology. At present, additional information concerning this Conference can be obtained from the General Chairman, Dr. James Aldrich, MS D461, P.O. Box 1663, Los Alamos National Laboratory, Los Alamos, NM 87545, USA.

Proceedings of the first three International Conferences on Basement Tectonics can be obtained by requesting an order form from S. Parker Gay, Jr., Applied Geophysics, Inc. 675 South 400 East, Salt Lake City, Utah 84111 or, Basement Tectonics Committee, Inc., P.O. Box 8568, Denver, CO 80201 USA. The three published volumes of this series now comprise over 1600 pages of text and figures devoted to some aspect of basement tectonics.

Robert A. Hodgson,
912 North Highland Ave.
Pittsburgh, PA 15206 U.S.A.